Downstate New York Rock Walks

Downstate New York
Rock Walks

An Explorer's Guide
to Amazing Boulders
and Rock Formations

C. Russell Dunn

Cover Credit: Bottom right to clockwise: Giant rock along Lake Sebago shoreline; Laura Petersen Balogh passes under the out-thrusting Hippo Rock while hiking the Long Path; Rock formation along Timp-Torne Trail. Photos by Dan Balogh.

Published by State University of New York Press, Albany

Printed in the United States of America

Excelsior Editions is an imprint of State University of New York Press

For information, contact State University of New York Press, Albany, NY
www.sunypress.edu

Library of Congress Cataloging-in-Publication Data

Name: Dunn, C. Russell, author | Title: Downstate New York Rock Walks / C. Russell Dunn, author.
Description: Albany : State University of New York Press, [2023] | Series: Excelsior Editions | Includes bibliographical references and index.
Identifiers: ISBN 9781438494692 (e-book) | ISBN 9781438494708 (paperback : alk. paper)
Further information is available at the Library of Congress.

10 9 8 7 6 5 4 3 2 1

CAUTION: Outdoor recreational activities are by their very nature potentially hazardous and contain risk. Please refer to section on **Caution: Safety Tips** for further details.

PRIVATE PROPERTY must always be respected! If you encounter posted land or no-trespassing signs while doing a hike, immediately turn around unless permission has first been obtained from the landowner.

CONTENTS

Westchester County

The Bronx

Nassau County

Manhattan

Queens

Staten Island (Richmond County)

Long Island (Suffolk County)

Harriman State Park and the Palisades

Harriman State Park: Northern Section

Harriman State Park: Southern Section

Bergen County

The Palisades

ILLUSTRATIONS

I1. The Hudson Highlands from Breakneck Mountain.
Photograph by the author.

INTRODUCTION

Downstate New York Rock Walks is a hiking guidebook that also doubles as a historical record of downstate New York's amazing boulders and rock formations.

If we had wanted to, we could have simply recounted the unique geological and cultural history of these incredible boulders and rock formations and left it at that. But in doing so we would have ignored one crucial element: People not only want to read about amazing boulders and rock formations, they want to get out and see them firsthand; hence, the reason for directions being included, where possible.

What you have before you is the culmination of untold hours of extensive library and online research in combination with technological advances available through Google Earth, MapQuest, and Topo software, all of which have allowed us to travel downstate, and into densely populated areas, without actually leaving our house in Albany (although, truth be told, we did get to actually visit a significant number of these sites).

Here, then, is an opportunity to hike to historic rocks, balanced rocks, perched rocks, rock monuments, talus caves, rock-shelters, tiny caves, massive glacial erratics, glacial potholes, and rock profiles—destinations other than your typical mountain summit, waterfall, and lake.

Although the directions given are pretty accurate thanks to MapQuest and Google Earth, there is always the chance we may have erred through misinformation or misunderstanding. For this reason, be prepared to make some allowances (but hopefully not many) where needed.

12. Approaching the entrance to Elbow Brush. Photograph by Dan Balogh.

Why Rocks?

We began thinking about writing a book on boulders and unusual rock formations of Downstate New York over ten years ago. That project, however, was momentarily set aside when a Massachusetts photographer named Christy Butler decided that we should do a rock-oriented hiking guidebook of Massachusetts, and so was born *Rockachusetts: An Explorer's Guide to Boulders and Rock Formations of Massachusetts* (2016).

It was while working on *Rockachusetts* that we both came to realize that we were doing something special, perhaps something that had never been done before. Despite many books having been published on rock climbing and bouldering, ours, we believed, was the first to feature boulders and rock formations as hiking destinations, each one unique in its own right. Since then, Christy has gone on to publish *Erratic Wandering: An Explorer's Hiking Guide to Vermont, New Hampshire & Maine* (an incredible undertaking by anyone's standards). Meanwhile, we have been busy at work on *Rambles to Remarkable Rocks: An Explorer's Guide to Amazing Boulders and Rock Formations of the Greater Capital Region, Catskills & Shawangunks; Boulders Beyond Belief: An Explorer's Hiking Guide to Boulders and Natural Rock Formations of the Adirondacks*, and *Rocks That Rock: An Explorer's Hiking Guide to Amazing Boulders and Rock Formations of Central & Western New York*.

The obvious question that inevitably comes to mind is: "Why write a book about rocks when everything around us is rock?" Truth be told, can there be anything more ordinary and commonplace than rock?

This argument would be unassailable if Earth was simply a mantle of unbroken, unvarying rock. It isn't—not by a long shot. The landscape before us has been immensely shaped and reshaped by plate tectonics, erosion, and repeating periods of glaciation. Enormous boulders have been picked up and moved incredible distances. Talus slopes have formed where sections of rock have been ripped off from cliff faces by glaciers or collapsed on their own accord. Rivers have created enormous chasms that continue to deepen. Swirling whirlpools, seizing stones and spinning them endlessly around, have augered their way into streambeds to create potholes. Enterable fissures have formed where the bedrock has literally split asunder like the shell of an egg cracking under pressure. Softer rock has been eroded out from under more durable rock to leave behind rock-shelters.

Yes, nature has been at work, and, as an artist, has sculpted some pretty amazing natural rock formations.

When it comes to downstate New York, we have been particularly impressed by just how many of the countless number of rocks are historically significant. Hundreds have been known about for centuries, many going back to the days when Native Americans used boulders and rock formations for meeting places and as refuges from the elements.

Until the last few centuries, large rocks served as natural points of navigation through wildernesses that were otherwise featureless. They also served as boundary markers for European settlers when lines of demarcation needed to be established for land ownership.

Many downstate rocks were widely publicized during the late nineteenth century and early twentieth century, their images reproduced through hundreds of thousands of postcards. You will see a number of these postcard reproductions scattered through the book.

Erratic Behavior

Downstate New York possesses a great many small-to-large erratics. Erratics, by definition, are rocks that are not native or indigenous to their surroundings. They have traveled from other regional areas, often over great distances. Early scientists were perplexed by the presence of erratics, confounded over what kind of force was powerful enough to move rocks the size of houses and as massive as freight cars over tens to hundreds of miles.

Native Americans, of course, had their own theory about erratics. Out-of-place rocks simply fell from the sky, perhaps dropped by the Great Spirit himself. As it turns out, the notion of rocks falling from the sky is not as fanciful as it may have once sounded. Earth, as we now know, is constantly being bombarded by meteorites, with some making it to the ground as meteors. Fortunately, most are burned up and disintegrate before reaching terra firma, which is why you are unlikely to ever come across a meteor in your travels. For this reason, we can safely rule out meteors for being the source of the thousands upon thousands of erratics that lay about. Are there any other theories?

The first scientific-sounding theory to explain the existence of erratics was called the Diluvial Theory. It drew inspiration from the Bible and postulated that it was the torrential deluge of waters from Noah's Great Flood that pushed and tumbled these huge rocks across the landscape. In this respect, the Diluvial Theory did make sense, for we all know the power of moving water and the tremendous force it can exert on objects encountered. The theory, however, only worked if you believed in a biblical, worldwide, Great Flood. It blithely ignored the fact that millions of huge erratics were inexplicably not present in the southern hemisphere as you would expect if a worldwide flood once raged. It also failed to answer how such enormous rocks managed to get up to the top of high mountains given the fact that fast-moving waters would push boulders around obstacles, such as hills and mountains, as opposed to carrying them up to the top.

In 1833, the Diluvial Theory was made slightly more palatable by bringing into play the role of icebergs. It was called the Ice-Rafted Theory. Now, in addition to Noah's Great Flood, it was postulated that huge icebergs broke loose from the arctic ice circle and, like rafts, transported boulders south over great distances, dropping them as the ice slowly melted. This theory both explained how boulders ended up on top of mountains (the mountains were under water), and why few large boulders are found in the southern hemisphere (the melting icebergs dropped their load before they could get there). The problem with this theory, however, was that it couldn't explain how boulders began their journey on top of the icebergs. That went against the natural order that ice forms on rocks—not vice versa.

It was up to Swiss-born Louis Agassiz to come up with a satisfactory theory, which he advanced in 1837. It was called the Glaciation Theory, and asserted that erratics, or what we now call glacial erratics, were swept up by stupendous glaciers. These mile-high glaciers bulldozed the landscape, moving rocks and earth southward, and, in the process, depositing millions of large rocks in their wake as the Earth warmed and the glaciers began to retreat northward. Since glaciers never reached the southern hemisphere, that explained why erratics were present only in the northern hemisphere.

Since Agassiz's time, there has never been any further debate about the matter (except for Creationists who believe that the Earth is only 6,000 years old and that there has been insufficient time for such momentous events to have happened).

We find it somewhat ironic, however, that in the end, it *was* water after all that moved rocks across the landscape—only, in this case, water in its solid form as ice.

Downstate New York 10,000 Years Ago

It's important to understand that during the last Ice Age, North America looked a great deal different than it does today. At that time, a huge portion of the ocean was locked up as ice, causing sea levels to drop over 200 feet and exposing vast tracts of land. Long Island was about twice as wide then as it is now. The Hudson River ran through a canyon whose walls were over 200 feet in height.

When the last Ice Age ended as glaciers retreated northward, a freshwater body of water called Lake Hudson momentarily formed. Its waters were temporarily dammed up by an earthen barrier that stretched between Brooklyn and Staten Island.

As far as Manhattan goes, there was little of it to see, just a portion of what we today call Midtown. Queens and the Bronx were essentially buried under Lake Flushing (where Flushing Meadow in Queens is now), and the East River (really just an arm of the Atlantic Ocean) that separated Manhattan from Brooklyn and Queens, had yet to come into existence.

It really was a different world back then as glaciers made their northward retreat!

How Big Is Big?

You might wonder how big a boulder can get. As it turns out, you don't have to look very far. The largest freestanding boulder on the planet is reputedly located right here in the United States. It is the seven-story high Giant Rock (34°19.970'N, 116°23.325'W) resting in the Mojave Desert near Landers, California. This is truly a humongous boulder. But large boulders aren't big just because of their size. They are massive—incredibly so. Double the size of a four-foot boulder to eight feet and its mass (weight) hasn't just changed by a factor of two. It has increased cubically, meaning that what started off at ten tons now weighs one thousand tons (10 x 10 x 10).

Largeness, then, really does matter when we talk about big boulders.

Balanced Rocks and Earthquakes

Large balanced rocks are astonishing when encountered, and serve to inspire one's imagination about the wonders of nature. But can they also serve a utilitarian function and be used to study past earthquakes? The answer to this very intriguing question is "yes." Paleoseismologists (scientists who study past earthquakes) are now hiking to large balanced and perched rocks to take pictures from multiple angles to create, through photogrammetry, a three-dimensional representation that can then be fed into a computer to calculate not only the mass, weight distribution, and balanced points of the rock, but what magnitude earthquake it would take to destabilize the rock.

William Menke, a geologist and seismologist with Columbia Climate School's Lamont-Doherty Earth Observatory in Palisades, New York, has made a number of trips into the nearby 47,500-acre Black Rock Forest to locate and take such measurements on large balanced rocks. Scientists are hopeful that the data generated will prove helpful in understanding our seismic past as well as creating future hazard models.

GPS Coordinates

Virtually all GPS coordinates listed in the book have been taken directly from Google Earth.

This means, then, that you may encounter minor discrepancies between your hand-held GPS unit and the Google Earth GPS coordinates, this being the result of Google Earth having to stitch together thousands of individual maps to create one world-encompassing map.

On the plus side, you can feed all the GPS coordinates for each chapter into your Google Earth program if you wish to reconnoiter the area in advance.

Delorme *New York State Atlas & Gazetteer* Coordinates

You will find that Delorme coordinates are ideal for providing an overview of the area that you are traveling through to reach the parking area or trailhead.

Be aware, however, that Delorme recently updated their *New York State Atlas & Gazetteer*, completely reversing the numerical order of their maps. For this reason, two sets of Delorme coordinates are provided to ensure that regardless of which edition you are using, you will be able to find your way with maximum efficiency.

Directions

Most of the directions have been created using various combinations of Google Earth, MapQuest, and Topo Mapware.

In one or two instances where navigating by MapQuest through a quagmire of streets and expressways was too complicated for a simple set of directions, sufficient information has been provided to get you to the general area, leaving you to figure out the rest using local maps.

In almost all instances, the directions start from a major intersection, such as "junction of Routes 9 & 311," and from there proceed to the trailhead parking.

In areas of high population density, like Manhattan, there are times when you may have to follow through on your own to find a parking space.

WOW Factor

A somewhat arbitrary "WOW Factor" number, ranging from 1 to 10, is given for each rock formation. The higher the number, the bigger the wow.

Photographing Rocks

The ideal always is to include a person in the photograph where possible. Doing so not only humanizes the photograph but provides important information regarding the size of the rock formation.

Secondly, most rocks tend to be in woodlands surrounded by trees. If possible, try to take as unobstructed a photograph of the boulder as possible. Trees and branches can get in the way and obscure the boulder or, minimally, act as a source of visual distraction.

Proper lighting also is a factor that cannot be disregarded. Since many boulders are located in deep woods, the light coming in can be severely restricted by trees, brush, and leaves. This fact needs to be taken into account when adjusting the setting on your camera.

On the other hand, if there is too much sunlight, the bright rays poking through branches and bushes can create a checkerboard pattern, once again making the image visually distracting. In this case, there may be little you can do to compensate except to wait for a momentary cloud to diffuse the sunlight.

This Is an Explorer's Guidebook

Once again, this is not just a hiker's guidebook, but also an explorer's guidebook. At times, you may need to rely upon your own ingenuity to locate a particular boulder or rock formation. However, what you are guaranteed to have is an adventure, and isn't half the fun just getting out into the great outdoors?

Regrettably, not all of the rocks described are necessarily accessible. In some cases, the boulder or rock formation may be on private land where visitors are not welcomed or where, minimally, permission from the landowner is required. We've included these rocks, not only for the sake of historical record (which is extremely important), but because there's always the possibility that the land may one day return to the public domain through the acquisition of lands from conservation-minded public and private agencies.

In other cases, the rock may simply no longer exist, destroyed either by man or by nature. This is particularly regretful when it is humans who are to blame.

Rocks that are nonexistent or currently inaccessible are listed as "historic."

CAUTION: SAFETY TIPS

Nature is inherently wild, unpredictable, and uncompromising. Outdoor recreational activities are by their very nature potentially hazardous and involve risk. All participants in such activities must assume the responsibility for their own actions and safety. No book can replace good judgment.

The outdoors is forever changing. The author and the publisher cannot be held responsible for inaccuracies, errors, or omissions; for any changes in the details of this publication; for the consequences of any reliance on the information contained herein; or for people's disregard for safety in the outdoors.

Remember: the destination is not the boulder, rock formation, or the mountain summit. The destination is home, and the goal is getting back there safely.

1. Always hike with two or more companions whenever possible. That way, if one person becomes incapacitated, one companion can go for help while the second stays close at hand to administer comfort and support.
2. Make it a practice of bringing along a day pack complete with emergency supplies, compass, whistle, flashlight, dry matches, rain gear, energy bars, extra layers, gorp, duct tape, lots of water (at least twenty-four ounces per person), mosquito repellent, emergency medical kit, sunblock, and a device for removing ticks.
3. Use sunblock when exposed to sunlight for extended periods of time and apply insect repellents as needed.
4. Hike with ankle-high boots—always!
5. Be cognizant of hypothermia (overcooling) and hyperthermia (overheating). Bring along extra layers when the temperature is cold, and drink plenty of water when the weather is hot and muggy.
6. Stay out of the woods during hunting season in areas where hunters are present.
7. Stay on trails unless you are proficient in orienteering.
8. Be aware that trails described in this book can become altered by blowdown, beaver dams, avalanches, mudslides, forest fires, and other natural disasters. Stay alert.
9. Always let someone know where you are going, when you will return, and what to do if you have not shown up by the designated time.
10. Leave early in the morning if you are undertaking a long hike, especially during the winter when daylight is at a premium.

11. Be mindful of ticks, which have become increasingly prevalent and virulent as their range expands. Check yourself thoroughly after every hike and remove any attached tick immediately.

So let's begin . . .

Downstate New York is huge, with many counties, boroughs, and islands. For this reason, particularly as a matter of convenience, we will use the Hudson River as a handy line of demarcation and divide the book into two sections: rocks on the east side of the Hudson River and rocks on the west side of the Hudson River.

With this said, we began our adventure, somewhat arbitrarily, starting with the east side of the Hudson River.

East Side of Hudson River

This section of the book covers all of Dutchess, Putnam, Westchester, Nassau, and Suffolk Counties; the New York City boroughs, including Manhattan and Staten Island; and all of Long Island.

Dutchess County

Dutchess County encompasses 825 square miles of land and is named for Mary of Modena, the Dutchess of York. The terrain is mostly hilly, with the Taconic Range to the northeast, and the Hudson Highlands to the southwest. Dutchess County is located in the mid-Hudson region of the Hudson Valley.

Stone Church Cave

1

Type of Formation: Cave
WOW Factor: 8
Location: Dover Plains (Dutchess County)
Tenth Edition, NYS Atlas & Gazetteer: p. 104, C1–2;
Earlier Edition, NYS Atlas & Gazetteer: p. 37, BC7
Parking GPS Coordinates: 41°44.330′N, 73°34.774′W
Stone Church Cave GPS Coordinates: 41°44.289′N, 73°35.324′W
Accessibility: 0.4-mile hike
Degree of Difficulty: Moderately easy

1. Inside Stone Church Cave. Photograph by the author.

Description

The Stone Church Cave is very impressive, and very much in keeping with what most people visualize when they bring to mind the image of a cave. Its vaulted entrance stands 30 feet high and is nearly as wide. At first glance, the cave almost looks man-made, much like a railroad tunnel, except that no train will ever emerge from it. The cavern extends back for 75 feet and then ends, but not before a 20-foot-high waterfall crashes down onto the bedrock of the floor.

A stupendous rock inside the cave, reaching almost halfway to the ceiling and quite possibly once a part of the ceiling that has dropped onto the floor, was reputedly named "The Pulpit" in 1908 by Richard Maher, author of *Historic Dover*, because of the rock's resemblance to an old-fashioned New England pulpit. It's entirely possible that it was through "The Pulpit" that the cave came to be named Stone Church Cave. According to Virginia Palmer, in her article "The Stone Church," the rock's "base is approximately twenty-five feet wide, and its height about thirty feet." Palmer goes on to mention that "a large boulder, oblong-shaped, stands upright in the rear of the cave." She also writes, "In the interior of the Stone Church, rock formations which some years ago were likened to pews and alters are no longer in evidence, or only slightly."

The cave was created by Stone Church Brook, a tributary of Ten Mile River. Upstream from the cave, the river has cut out a fairly deep, narrow gorge with multiple cascades.

History

Historical accounts tell us that Sassacus, a Mashantucket Pequot chief, and a band of his warriors took refuge in the cave in the 1600s during King Philip's War, hoping to evade capture and execution by the British and Mohawks. After leaving the cave, however, they were caught by the Mohawks and summarily dispatched.

Stone Church Cave became a popular tourist mecca in the 1830s. Droves of visitors from New York City and adjacent states would descend on Dover Plains to visit the cave. It was celebrated in much the same way that Howe Caverns in Schoharie County was. Weddings were even performed in the cave, which is another possible explanation for how the cave came to be named.

Stone Church Cave's prominence rose to even greater heights after it was rendered onto canvas by several Hudson River School painters (one of them being Asher Durand) and featured by Benson Lossing in *Dover Stone Church and the History of Dutchess County.*

At one time, the waters from Stone Church Brook were impounded by a cement dam and used as a water supply for the residents of Dover Plains. By 1950, virtually all traces of the cement dam were gone.

The land was acquired by the Dutchess County Open Space and Farmland Protection Fund, Friends of Dover Stone Church, and the Dutchess County Land Conservancy in 2002. This acquisition included not only the cave, but 58 acres of surrounding land as well as a conservation easement from Route 22 to the cave. Improvements were later made on the trail leading to the cave.

Dover Plains was named after the white chalk cliffs of Dover, England.

Directions

From Amenia (junction of Routes 22, 343 & 44), drive south on Route 22 for 9.0 miles to Dover Plains.

From Wingdale (junction of Routes 22 & 55), drive north on Route 22 for 7.0 miles to Dover Plains.

Park at the Dover Elementary School or designated area should a new parking area be created.

Look for a New York State historic sign for Stone Church Cave on the west side of Route 22, nearly opposite Mill Street. Follow the driveway uphill for 100 feet to reach the trailhead (41°44.383' N 73°34.924' W). Take note that parking here is reserved for homeowners in the adjacent houses.

From the trailhead, follow a flight of modern, railed, stone block steps down the side of a small hill to the valley floor, across an open field through a grove of old trees into the woods, and then next to Stone Church Brook on your left. Soon, a footbridge takes you across to the opposite side of the stream.

By 0.3 mile, the path enters into a gorge. Slabs of rock provide ample footing. In another 0.1 mile or less you will reach the entrance to Stone Church Cave. If the stream is not running fast, there are sufficient slabs of rock to enable you to rock-hop into the interior of the cave.

Resources

Russell Dunn, *Hudson Valley Waterfall Guide* (Hensonville, NY: Black Dome Press, 2005), 204–206.

"Dover Stone Church Cave – Scenic Hudson." https://www.scenichudson.org/explore-the-valley/outdoor-adventures/adventure/dover-stone-church/.
"Dover Stone Church Cave Visitor's Guide." https://www.oblongland.org/pdf/stone_church_dover.pdf.
Wallace Bruce, *The Hudson: Three Centuries of History, Romance, and Invention*, Centennial Edition (New York: Walking News, 1982), 127.
Patricia Edwards Clyne, *Caves for Kids in Historic New York* (Monroe, NY: Library Research Associates, 1980), 55–62.
Thomas Sweet Lossing, *My Heart Goes Home: A Hudson Valley Memoir* (Fleischmanns, NY: Purple Mountain Press, 1997), 107–108.
Clay Perry, *Underground Empire: Wonders and Tales of New York Caves* (New York: Stephen Daye Press, 1948), 130–131. A spectacular photograph of the cave can be seen on an insert between pages 54 and 55.
Joyce C. Ghee and Joan Spence, *Harlem Valley Pathways: Images of America* (Charleston, SC: Arcadia Publishing, 1998). A photograph of the cave entrance is displayed on page 33. A caption accompanying the photograph reads, "Magnetite in the rock fools compasses and ancient petroglyphs confound archaeologists." Are there actually petroglyphs at Stone Church Cave?
Virginia Palmer, "The Stone Church," *Year Book: Dutchess County Historical Society*, vol. 33 (1948), 45–49.
Frank Hasbrouck, *The History of Dutchess County, New York* (Poughkeepsie, NY: S. A. Mattheu, 1909). A photo of the cave can be seen on an insert between pages 282 and 283.
Russell Dunn and Barbara Delaney, *Paths to the Past: History Hikes through the Hudson River Valley, Catskills, Berkshires, Taconics, Saratoga & Capital Region* (Catskill, NY: Black Dome Press, 2021), 45–48.

Indian Rock 2

Margaret Lewis Norrie State Park: Esopus Island

Type of Formation: Petroglyph/Megalith
WOW Factor: 2–3
Location: Staatsburg (Dutchess County)
Tenth Edition, NYS Atlas & Gazetteer: p. 103, A7;
Earlier Edition, NYS Atlas & Gazetteer: p. 36, B4
Parking GPS Coordinates: 41°49.946′N, 73°56.459′W
Esopus Island GPS Coordinates: 41°49.506′N, 73°56.858′W
Accessibility: 0.4-mile trek by water
Degree of Difficulty: Easy by boat
Additional Information: Norrie Point Environmental Center, 9 Old Post Road, Staatsburg, NY

Description

In *Hudson Valley Trails and Tales*, Patricia Edwards Clyne writes about a "mysterious carved boulder in the [Hudson] river . . . known as Indian Rock." It is located on Esopus Island.

Esopus Island is a very narrow, 1,500-foot-long island, 200 feet across at its widest. It contains campsites, picnic areas, trails, and fishing access spots. One writer has likened the island's shape to that of a "great stranded and petrified whale." Most of the island's shoreline is exposed bedrock, which would seem to make locating the petroglyphs (images caved into rocks) and megaliths (large stones used to construct prehistoric structures) fairly difficult due to the island's extensive perimeter. Fortunately, one of our sources has narrowed the search down substantially, asserting that the Native American handiwork is on the east side of the island.

There is a small beach on the southeast side of the island, and shoals at its north end. We suspect most people dock at the beach area.

Located 1.4 miles south of Esopus Island is 0.1-mile-long Bolles Island, with a private residence at its south end.

History

The Lenape Native Americans made use of Esopus Island and, while on it, fashioned carvings and quasi-megaliths.

According to legend, a Jesuit missionary was killed on the island by the Lenape for reasons that remain unclear, but which we suspect might have been due to religious zealotry.

In 1777, the British fleet temporarily landed on the island before resuming course to attack and destroy Kingston.

During the nineteenth century, the island was part of Robert L. Pell's 1,200-acre estate and was then known as Pell Island.

By coincidence, there is also an Indian Rock (a cliff) in the same general area on the opposite side of the Hudson River, slightly northwest of Esopus Island. It can be seen clearly on mapcarta.com/22011990 (41°49.972'N, 73°57.344'W). We doubt, however, that confusion ever arises when people wish to delineate the two Indian Rocks.

Directions

From Staatsburg (junction of Routes 9 & 37/North Cross Road)), drive southwest on Route 9 for 0.5 mile and turn right onto Margaret Norrie State Park Road. At a fork, bear left to reach the Norrie Point Environmental Center, 1.0 mile from Route 9.

From the Norrie Point Environmental Center launch site (41°49.903'N, 73°56.531'W), head southwest on the Hudson River for <0.4 mile to reach the north tip of Esopus Island. The island can be circumnavigated in 0.9 mile, allowing you to choose a suitable landing spot as well as to check out sites to explore.

Resources

Patricia Edwards Clyne, *Hudson Valley Trails and Tales* (Woodstock, NY: Overlook Press, 1990), 258

Stanley Wilcox and H. W. Van Loan, *The Hudson from Troy to the Battery* (Philmont, NY: Riverview Publishing, 2011), 60. The authors tell the tale of Aleister Crowley, an astrologer and occultist, who is said to have spent forty days and nights on the island in 1941 while exploring deeper levels of consciousness.

"Esopus Island Has Become the Hudson Valley's Great Mystery:" https://www.thetravel.com/esophus-island-great-mystery-history-hudson-valley/.

Waryas Park Rocks and Kaal Rock 3

Waryas Park

Type of Formation: Strange Rock Sculpture; Historic Rock Bluff
WOW Factor: 2–3
Location: Poughkeepsie (Dutchess County)
Tenth Edition, NYS Atlas & Gazetteer: p. 103, C7
Earlier Edition, NYS Atlas & Gazetteer: p. 36, C4
Parking GPS Coordinates: *Rock Sculpture*: 41°42.333′N, 73°56.405′W; *Kaal Rock Park*: 41°42.185′N, 73°56.451′W
Destination GPS Coordinates: *Rock Sculpture*: 41°42.322′N, 73°56.415′W; *Kaal Rock*: 41°42.229′N, 73°56.456′W
Accessibility: *Rock Sculpture*: 20-foot walk; *Kaal Rock*: >0.1-mile trek uphill
Degree of Difficulty: *Rock Sculpture*: Easy; *Kaal Rock*: Moderately easy

---

Description

Waryas Park Rocks consist of three moderately large, artificially created rocks. Although it's not evident at first, these three separated rocks actually form one large, extended piece of artwork—the tail, back, and head of an enormous 80-foot-long sperm whale. The tail is about 7 feet in height, the head 4 feet.

Harvey K. Flad and Clyde Griffin, in *Main Street to Mainframes: Landscape and Social Change in Poughkeepsie*, write that "The sculptures were built out of steel rebar and cement and covered by ceramic tile mosaics."

Just 0.3 mile upriver looms the 1.3-mile-long bridge called Walkway Over the Hudson—the centerpiece of the Walkway Over the Hudson State Historic Park. It is said to be the longest elevated pedestrian bridge in the world.

Just south of the park, >0.1 mile upriver from the 3,000-foot-long Mid-Hudson Bridge is Kaal Rock—a 50-foot-high bluff that overlooks the Hudson River and 6.6-acre Kaal Rock Park. Karl Rock once served

2. Whale Rock Sculpture at Waryas Park. Photograph by the author.

as a vital reference point for sailors navigating the river. Ernest Ingersoll, in *Handy Guide to the Hudson River and Catskill Mountains*, writes that "tradition says the early burghers of the town used to sit [on Kaal Rock], and hail the sloops for news as they drifted by."

History

Waryas Park encompasses 9 acres of green space and was named after former Poughkeepsie mayor Victor C. Waryas, who served from 1960 to 1964.

The sperm whale art piece was created by Cragsmoor-based artist Judy Sigunick in 2002. The subject chosen was not done haphazardly. Poughkeepsie at one time was a vibrant port, with two whaling companies operating from what is now Waryas Park.

Kaal Rock, aka Caul Rock and Call Rock, is Dutch for "Bald Rock."

Directions

WARYAS PARK

Traveling north through Poughkeepsie along Route 9, get off at the Main Street exit. When you come to Main Street, turn left and drive 0.3 mile west to reach Waryas Park.

Traveling south on Route 9, get off at the Laurel Street exit, turn right onto Laurel Street, and head west for 0.05 mile to Rinaldi Boulevard. Turn right onto Rinaldi Boulevard and drive north for over 0.3 mile. Finally, turn left onto Main Street and proceed west for 0.1 mile to reach the park.

Park in the cul-de-sac. The cement sculptures are southwest of the cul-de-sac, a mere 50 feet away.

Kaal Rock is the wooded precipice beyond the south end of Waryas Park, lying between Waryas Park and Kaal Rock Park. To reach Kaal Rock, walk uphill from the sculptures, first heading east, then south, as you follow along the edge of the woods for 0.1 mile. When the hill levels off, you will be at the cul-de-sac at the end of Long Street. A path from here to your right leads into the woods and up to the highest point of Kaal Rock in 0.1 mile. This vertical precipice is an impressive overlook by anyone's standards. To your left (south) is the Mid-Hudson Bridge and directly below, also south, is Kaal Rock Park (which can be reached by following a rocky path downhill—a trek that is obviously not for everyone).

KAAL ROCK PARK

Leaving Waryas Park, turn right onto Rinaldi Boulevard and drive south for 0.2 mile. Bear right onto Gerald Drive and head southeast for 0.2 mile. When you come to Hendryck Drive, turn right and proceed northwest for 0.2 mile, passing under the Mid-Hudson Bridge and parking at the north end of the Park. The south shoulder of Kaal Rock is directly in front of you as you look north.

Resources

Joyce C. Ghee and Joan Spence, *Poughkeepsie Halfway up the Hudson: Images of America* (Charleston, SC: Arcadia Publishing, 1997), 29.

Harvey K. Flad and Clyde Griffin, *Main Street to Mainframes: Landscape and Social Change in Poughkeepsie* (Albany, NY: State University of New York Press, 2009), 227.

"Where in the Hudson Valley Contest: 'Wayward Whale' Park Sculpture." https://hvmag.com/Hudson-Valley-Magazine/June-2014/Where-in-the-Hudson-Valley-Contest-Wayward-Whale-Park-Sculpture.

Ernest Ingersoll, *Handy Guide to the Hudson River and Catskill Mountains* (Astoria, NY: J. C. & A. L. Fawcett, 1989; reprint of 1910 book), 132.

"Victor C. Waryas Memorial Park Historical Marker." https://www.hmdb.org/m.asp?m= 37868.

Lovers Leap (Historic) 4

Poughkeepsie Rural Cemetery

Type of Formation: Historic Bluff
WOW Factor: 3
Location: Poughkeepsie (Dutchess County)
Tenth Edition, NYS Atlas & Gazetteer: p. 103, C7;
Earlier Edition, NYS Atlas & Gazetteer: p. 36, C4
Lovers Leap GPS Coordinates: 41°40.818′N, 73°56.320′W (estimated)
Accessibility: Inaccessible
Additional Information: Poughkeepsie Rural Cemetery,
342 South Avenue, Poughkeepsie, NY 12601.

Description

Lovers Leap is a high bluff that overlooks the Hudson River and the Amtrak railroad. It is located along the southwest border of the Poughkeepsie Rural Cemetery—a 165-acre graveyard that dates to 1853.

History

A number of legends are associated with Lovers Leap, all involving one or two Native Americans or Victorians who commit suicide, generally out of unrequited or forbidden love. There is even one tale of a man pushing a woman off the bluff and then jumping to his own death.

These are very common stories, virtually all mythical, and associated with many high cliffs and large waterfalls in New York State.

In *Poughkeepsie Halfway up the Hudson: Images of America*, a photograph is shown of a well-dressed Victorian man sitting in a summerhouse (gazebo) by Lovers Leap. Access to the site has clearly changed over the centuries.

The private mausoleum (41°40.809′N, 73°56.295′W) next to Lovers Leap was erected in 1932 by Dr. Emile Alfred Muller for his beloved housekeeper, Alice Whittier.

Directions

From Poughkeepsie (at the point where Route 9 passes under the Mid-Hudson Bridge), drive south on Route 9 for >1.1 miles. Veer right onto Old South Road, which parallels Route 9, and then right into the entrance to the Poughkeepsie Rural Cemetery after 0.2 mile.

Drive southwest across a landscape of rolling hills to the north end of a pretty, 0.1-mile-long, unnamed pond. Take note of an interesting 8–10-foot rock outcropping by the northeast corner of the pond.

This is as far as you can go. Although a dirt road leads southwest for 0.3 mile through an isolated section of the cemetery to a mausoleum, and then from there <200 feet to the Lovers Leap precipice, the mausoleum and road are privately owned, and visitors are not allowed.

Resources

Joyce C. Ghee and Joan Spence, *Poughkeepsie Halfway up the Hudson: Images of America* (Charleston, SC: Arcadia Publishing, 1997), 94.

"Lover's Leap from the Poughkeepsie Rural Cemetery." https://riverletters.blogspot.com/2010/04/lovers-leap-from-poughkeepsie-rural.html. This blog provides background information about the mausoleum at the Poughkeepsie Rural Cemetery.

"Whittier Mausoleum - Poughkeepsie (NY) Rural Cemetery." https://waymarking.com/waymarks/WM9FM7_Whittier_Mausoleum_Poughkeepsie_NY_Rural_Cemetery.

Poughkeepsie Standing Stone 5

Type of Formation: Unique Stone
WOW Factor: 2
Location: Wappinger Falls (Dutchess County)
Tenth Edition, NYS Atlas & Gazetteer: p. 103, D7
Earlier Edition, NYS Atlas & Gazetteer: p. 36, CD4
Poughkeepsie Standing Stone GPS Coordinates: 41°36.620′N, 73°55.896′W
Accessibility: Roadside
Degree of Difficulty: Easy

Description

The Poughkeepsie Standing Stone is a rectangular, 4-foot-high slab of limestone jutting out of the ground at an inclined angle.

History

The Poughkeepsie Standing Stone is believed to have been shaped by human hands rather than by weathering. It is not a natural stone then. The rock is called a menhir, which is a French term for a large, upright, standing stone of variable size, its shape, generally uneven, often tapering near the top.

Laura M. Lane, who once lived across from the Standing Stone, never gave the rock much thought until her dad, an engineer, did some plotting, and figured out that 17 feet of the rock's length would have to be buried underground in order for the rock to sit as it does.

Whether 17 feet of this rock actually lie buried underground or not, it serves to illustrate that the Standing Stone has been a point of curiosity for many years. Joyce C. Ghee and Joan Spence, in *Poughkeepsie Halfway up the Hudson: Images of America*, even go so far as to say that the stone has been here "since before recorded history."

3. Poughkeepsie Standing Stone. Photograph by the author.

David Beck, a park naturalist at Bowden Park, recounted to us a story he was told by a local historian who claimed that Native Americans used the stone to sight on a sacred ceremonial site that lay uphill where the Mount Alvernia Retreat Center (Order of Friars Minor) is now located.

However, others believe that an even deeper meaning lies within the stone. On one of his Spirits in Stone: The Secrets of Megalithic America websites, author Glenn Kreisberg recounts a story in Sal Trento's 1978 book, *The Search for Lost America*, in which Trento and his group, after determining that the Standing Stone is tilted in the direction of 202°, follow that course for less than two miles to the mouth of Wappinger Creek, from where they can readily see Danskammer Point [see chapter on "Devil's Dance Hall"] across the river, another supposedly mystical site. And so, a preternatural connection is made between these two disparate sites, if you believe this kind of thing.

Directions

From northwest of Spackenkill (junction of Routes 9 & 113/Spackenkill Road), drive south on Route 9 for 1.7 miles. Turn right onto Sheafe Road and continue southwest for another 2.0 miles. Finally, turn left onto Delavergne Avenue and drive east for 0.2 mile. The stone is at the intersection of Delavergne Avenue and Oakwood Drive, virtually opposite the road leading up to the Mount Alvernia Retreat Center.

Resources

Joyce C. Ghee and Joan Spence, *Poughkeepsie Halfway up the Hudson: Images of America* (Charleston, SC: Arcadia Publishing, 1997), 86.

"Ancient Astronomy along the Hudson River in New York State." https://grahamhancock.com/kreisbergg7.

"Mysterious Hudson Valley Stone Sites." https://www.facebook.com/MysteriousHudsonValleyStoneSites/posts/863182113847443/. This site contains a photograph of the standing stone.

Bowdoin Park Rockledge Shelters 6

Bowdoin Park

Type of Formation: Rock-Shelter
WOW Factor: *North Rockledge Shelter*: 4–5; *South Rockledge Shelter*: 4
Location: Wappinger Falls (Dutchess County)
Tenth Edition, NYS Atlas & Gazetteer: p. 103, D7
Earlier Edition, NYS Atlas & Gazetteer: p. 36, CD4
Main Parking GPS Coordinates: 41°36.144'N, 73°56.452'W
Destination GPS Coordinates: *North Rockledge Shelter*: 41°36.332'N, 73°56.455'W; *South Rockledge Shelter*: 41°35.835'N, 73°56.371'W
Accessibility: *North Rockledge Shelter*: 0.2-mile walk, followed by a 100-foot uphill trek; *South Rockledge Shelter*: 0.4-mile hike to top of bluff, followed by 200-foot bushwhack down to the rocky base of escarpment
Degree of Difficulty: *North Rockledge Shelter*: Moderately easy; *South Rockledge Shelter*: Moderately easy to bluff overlook; moderately difficult to base of bluff
Additional Information: Dutchess County Bowdoin Park, 85 Sheafe Road, Wappinger Falls, NY 12590; (845) 298-4600
Trail Map: dutchesstourism.com/PDF/brochure-rack/Bowdoin_Trail_Map_Brochure_5_24_10.pdf

---

Description

Bowdoin Park contains two historically significant rockledge shelters. Both have formed at the base of a rocky escarpment, and both have downslopes that lead from their base to the floor of the valley.

The North Rockledge Shelter, roughly 6–7 feet high, lies at the bottom of a 30-foot-highcliff . Four feet of debris and earth had to be removed from the entrance in order to restore the shallow cavity to its original state when used by Native Americans thousands of years ago. It's possible that

4. Park naturalist David Beck stands next to the North Rock-Shelter. Photograph by the author.

the rock-shelter may have been wider back then, a fact that can only be determined if further excavation is undertaken. It's also possible that the overhang may have once extended out farther from the cliff, but no one knows for certain.

By good fortune, David Beck, park naturalist, accompanied us to the rock-shelter and proved very helpful in translating what we were seeing into what the site was like centuries—even millenniums—ago. At that time, the land was not as devoid of sustenance as it is today. Game was more plentiful, and chestnut trees (which were literally everywhere until wiped out by the chestnut blight in the early 1900s) provided plenty of nourishment.

It's very likely that Native Americans lived much closer to the river during the warmer months, retreating to the rock-shelter seasonally when conditions by the river became less hospitable.

Part of the rock-shelter is presently covered with a wooden frame thatched with reed grass in order to show visitors how Native Americans created an enclosure to keep out the harsh weather.

The South Rockledge Shelter, at the opposite end of Bowdoin Park, is located at the bottom of a 40-foot-high bluff, with considerably more rock face showing, as well as more of an overhang than its northern counterpart. The rockledge shelter evidently gets fewer visitors, for no discernable path leads down to it.

In their book *Poughkeepsie Halfway up the Hudson: Images of America*, Joyce C. Ghee and Joan Spence mention that, historically, the rock-shelters were easily accessible from the Hudson River, and provided inhabitants with an unobstructed view of the river for defensive purposes. The view today isn't quite as unobstructed.

History

Bowdoin Park encompasses 301 acres of lands and contains roughly 5 miles of trails. In the 1920s, the Children's Aid Society received the Bowdoin estate as a bequest. The land passed on to the county in the 1960s and opened to the public in 1975.

It is believed that a Native American village may have occupied the site where the soccer fields are located today.

Directions

From northwest of Spackenkill (junction of Routes 9 & 113/Spackenkill Road), drive south on Route 9 for 1.7 miles. Turn right onto Sheafe Road and continue southwest for another 2.6 miles. Then turn right into the entrance of Bowdoin Park and follow the main road downhill for 0.2 mile to a mid-level parking area.

NORTH ROCKLEDGE SHELTER

From the parking area, walk north to the four-way intersection you just turned left at. Cross the road and continue past a red-colored sign that says "Private. No Vehicles Allowed," following a dirt road north for 0.2 mile. At the time of year when trees are leafless, you will notice an old road to your right that gradually gets closer and closer to the dirt road until it merges. At that point, you will see a faint path to your right that leads uphill in 150 feet to the rockledge shelter.

SOUTH ROCKLEDGE SHELTER

Walk east uphill from the parking area for less than 100 feet to reach the trailhead. Follow the white-blazed trail south for 0.2 mile. At a fork (across from a bench), bear right and continue south on the now yellow-blazed trail for another 0.2 mile. When you come to a bare, rocky prominence overlooking the valley, you are directly above the rockledge shelter. Backtrack a hundred feet or so, and then bushwhack southwest down the slope to reach the bottom of the escarpment and the rock-shelter. Expect a fairly demanding descent over blowdown and loose rock. For most visitors, the view from the rocky bluff above the rock-shelter will prove sufficient.

Resources

Joyce C. Ghee and Joan Spence, *Poughkeepsie Halfway up the Hudson: Images of America* (Charleston, SC: Arcadia Publishing, 1997), 86.

Joyce C. Ghee and Joan Spence, *Poughkeepsie 1898–1998. A Century of Change: Images of America* (Charleston, SC: Arcadia Publishing, 1999), 75.

Harvey K. Flad and Clyde Griffin, *Main Street to Mainframes: Landscape and Social Change in Poughkeepsie* (Albany, NY: State University of New York Press, 2009), 166.

Jeffrey Perls, *Paths along the Hudson: A Guide to Walking and Biking* (New Brunswick, NJ: Rutgers University Press, 2001), 324–325.

"Bowdoin Park – Dutchess County Government." https://www.dutchessny.gov/Departments/Parks/Bowdoin-Park.htm.

Nuclear Lake Rocks

7

Nuclear Lake

Type of Formation: Medium-Sized Rock; Talus Slope
WOW Factor: 3
Location: West Pawling (Dutchess County)
Tenth Edition, NYS Atlas & Gazetteer: p. 104, D1
Earlier Edition, NYS Atlas & Gazetteer: p. 37, CD6–7
Parking GPS Coordinates: 41°35.149′N, 73°39.174′W
Nuclear Lake Rocks GPS Coordinates: *One set of rocks*: 41°35.677′N, 73°38.779′W
Accessibility: 0.7-mile hike to Nuclear Lake; 0.7-mile hike along west side of lake along Appalachian Trail; 1.1-mile hike along east side of lake following Nuclear Lake Loop Trail
Degree of Difficulty: Moderate

Description

Nuclear Lake provides an interesting destination to beautiful lake scenery, rocky outcrops, talus fields, and a cluster of medium-sized boulders. Although we have seen photographs of hikers bouldering on some sizeable rocks at the lake, we were not able to locate these larger rocks when we circumnavigated the lake—only the medium-sized ones. Perhaps you will have better success.

History

Nuclear Lake got its name when an experimental nuclear fuel research lab, established on the southwest shore of the lake in 1955, exploded in 1972, scattering a significant amount of bomb-grade plutonium over the nearby woods and water. Fortunately, it was a chemical explosion, not a nuclear one. To be sure, this wasn't the first incident. In 1971, a rubber stopper came off a plutonian powder container, contaminating the lab room.

Although hikers may jest in earnest about the risk of encountering traces of plutonium around the lake, the area was tested in 1994 by the Nuclear Regulatory Commission and found to be perfectly safe.

Nuclear Lake is a 55-acre, man-made body of water, formed by an earthen dam at the south end. It was at this end where a series of buildings doing experimental research once stood.

Directions

From Poughquag (junction of Routes 55/Freedom Plains Road & 216), proceed southeast on Route 55/Freedom Plains Road for 1.8 miles. Turn left onto Old Route 55 and then an immediate left into a parking area specifically for the Nuclear Lake hike.

From Pawling (junction of Routes 55 & 9), drive northwest on Route 55 for 4.8 miles. Turn right onto Old Route 55, and then immediately left into a parking area for the Nuclear Lake hike.

From the parking area, follow a dirt road northeast for >0.7 mile to reach the earthen dam at the south end of Nuclear Lake. Along the way, <0.1 mile before reaching the lake, take note of the yellow-blazed Nuclear Lake Trail Loop that enters on your right. If you completely circle the lake, you will exit at this point.

Follow a continuation of the old road which is soon marked by the white-blazed Appalachian Trail (AT). After >0.7 mile, the north end of the lake and a junction is reached. Were you to continue north on the AT, you would eventually come to Cat Rock at 3.0 miles, a panoramic, east-facing overlook. According to the *Guide to the Appalachian Trail in New York and New Jersey*, Cat Rock is "another conglomerate outcropping, with top leveled by glacial action."

Turn right instead and follow the yellow-blazed Nuclear Lake Trail Loop south along the east side of the lake. This is where a number of outcroppings can be found as well as several talus slopes and medium-sized rocks. Eventually, you will return to the road that you started on, 0.1 mile before the lake.

Resources

"Nuclear Lake." https://hikethehudsonvalley.com/hikes/nuclear-lake.
"Nuclear Lake." https://hudsonvalleygeologist.blogspot.com/2014/12/nuclear-lake.html.
"How To Visit New York's Nuclear Lake (Yes, This Exists)." https://scoutingny.com/how-to-visit-new-yorks-nuclear-lake.
"NY 55 to Connecticut State Line." https://cnyhiking.com/ATinNY-HammersleyRidge.htm.
New York–New Jersey Trail Conference, *Guide to the Appalachian Trail in New York and New Jersey*, 9th ed. (Harpers Ferry, WV: Appalachian Trail Conference, 1983), 93.

Putnam County

Putnam County encompasses 246 square miles of land and is generally hilly. The county was named after Israel Putnam, a general in the American Revolutionary War who also fought in the French and Indian War. It is located in the lower Hudson River Valley.

Split Rock

8

Hudson Highlands State Park

Type of Formation: Split Rock; Large Boulder
WOW Factor: 6
Location: Nelsonville (Putnam County)
Tenth Edition, NYS Atlas & Gazetteer: p. 108, A2
Earlier Edition, NYS Atlas & Gazetteer: p. 32, A4
Parking GPS Coordinates: 41°26.021′N, 73°56.215′W
Destination GPS Coordinates: *Split Rock*: 41°26.318′N, 73°56.250′W;
Large Boulder: 41°26.484′N, 73°56.145′W
Accessibility: *Split Rock*: 0.3-mile hike;
Large Boulder: 0.2-mile bushwhack from Split Rock
Degree of Difficulty: Moderate
Additional Information: Trail map available at East Hudson Trails:
Trail Map 102, 11th edition
Trail Map: parks.ny.gov/documents/parks/HudsonHighlandsTrailMapNorth.pdf

---

Description

Split Rock is a large, 10-foot-high boulder that has fractured into two, 8–10-foot-long pieces that lie separated by a space of two or more feet.

The second large boulder, most likely visited only infrequently, requires a short bushwhack from the Lone Star Trail to reach. It is quite visible on Google Earth, which is where we first noticed it.

History

The Hudson Highland State Park contains nearly 8,000 acres of wilderness, with parcels extending from Annsville Creek in Peekskill north to Dennings Point in Beacon. One of its most exciting attractions is Breakneck Ridge, which *Newsweek* rated as one of the ten best day hikes in America.

5. Ariel Schwartz pauses for a moment at Split Rock.
Photograph by Daniel Chazin.

Interestingly, Breakneck Mountain, which is part of Breakneck Ridge, was earlier known by a different name. According to Ernest Ingersoll, in *Handy Guide to the Hudson River and Catskill Mountains*, "A century ago [that would be in the 1800s] it was known as *The Turk's Face*, owing to a remarkable image of a human countenance, formed by projecting rocks on the south side, where now a purplish wall of bare rock testifies to the ravages of stone-quarrying; but his was long ago tumbled down by the operations of blasting."

The Split Rock Trail is a short connector trail between the Lone Star Trail and the Nelsonville Trail.

Directions

From McKeel Corners (junction of Routes 301 & 9), drive southwest on Route 301 for 1.5 miles. Turn sharply right onto Route 10/Fishkill Road and drive northeast for 0.4 mile. Park in a small area on your left.

SPLIT ROCK

Head north on the blue-marked Lone Star Trail, an old woods road, for >0.3 mile. You will see the large split rock to your left at the point where the red-marked Split Rock Trail comes in on your left.

LARGE BOULDER

From Split Rock, hike north, then slightly east for 0.2 mile, a comparatively short bushwhack.

Resources

New York–New Jersey Trail Conference, *New York Walk Book*, 6th ed. (New York: New York–New Jersey Trail Conference, 1998), 148.

Ernest Ingersoll, *Handy Guide to the Hudson River and Catskill Mountains* (Astoria, NY: J. C. & A. L. Fawcett, 1989; reprint of 1910 book), 111.

Goose Rocks

9

Oscawana Lake

Type of Formation: Large Boulder
WOW Factor: 4
Location: Oscawana Lake (Putnam County)
Tenth Edition, NYS Atlas & Gazetteer: p. 108, B3
Earlier Edition, NYS Atlas & Gazetteer: p. 33, A5
Oscawana Marina Parking GPS Coordinates: 41°23.318′N, 73°50.851′W
Goose Rocks GPS Coordinates: 41°23.981′N, 73°51.021′W
Accessibility: 0.7-mile trek on water
Degree of Difficulty: Easy by boat
Additional Information: Oscawana Marina, 96 Dunderberg Road, Putnam Valley, NY 10579; Oscawana Lake Civic Association (LOCA): lakeoscawana.org

Description

Goose Rocks consist of a grouping of large boulders near the upper/middle section of 2.0-mile-long, 386-acre Oscawana Lake. Geologists believe that the rocks were dislodged and carried down by glaciers from the surrounding mountains. Two of the bigger rocks are roughly 30 feet long.

History

Most of Oscawana Lake is developed except for the west side which abuts the Clarence Fahnestock Memorial State Park.

It seems probable that the rocks were named after flocks of geese which use the rocks for refuge.

Oscawana is the name of a Native American personage who was involved in early land transactions.

Baseball icon Babe Ruth is reported to have spent time at the lake in the 1930s, his favorite spot being The Casino. Actor Roy Schneider also enjoyed time at the lake.

Directions

From Crofts Corner (junction of Routes 20/Oscawana Road & 22/Church Road), drive northeast on Route 20/Oscawana Road for 1.8 miles. Turn left onto Dunderberg Road and proceed north for >0.4 mile. The Oscawana Marina is on your right.

From the Marina, the trip to Goose Rocks is 0.7 mile by boat or kayak.

Resources

New York–New Jersey Trail Conference, *New York Walk Book*, 6th ed. (New York: New York–New Jersey Trail Conference, 1998). A line drawing of the rocks can be seen on page 38.

"Our Lake – Lake Oscawana." https://www.lakeoscawana.org/our-lake

6. *(opposite)* Goose Rocks jutting out of Oscawana Lake.
Antique postcard, public domain.

Hawk Rock and Balanced Rock 10

New York City Environmental Protection Land

Type of Formation: Large Boulder; Balanced Rock
WOW Factor: *Hawk Rock*: 6–7; *Balanced Rock*: Not determined
Location: Kenwood Lake (Putnam County)
Tenth Edition, NYS Atlas & Gazetteer: p. 108, A5
Earlier Edition, NYS Atlas & Gazetteer: p. 33, A6
Parking GPS Coordinates: 41°29.016′N, 73°42.054′W
Destination GPS Coordinates: *Hawk Rock*: 41°27.842′N, 73°41.642′W (estimated); *Balanced Rock*: Not determined
Accessibility: *Balanced Rock*: 1.5-mile hike; *Hawk Rock*: 1.7-mile hike
Degree of Difficulty: Moderate
Additional Information: Trail map: kentcac.info/wp/wp-content/uploads/2015/07/Hawk-Rock-Brochure-v3.pdf.
Note: An access permit from the Department of Environmental Protection is required in order to undertake this hike. A permit can be obtained at nyc.gov/html/dep/.

Description

The only description we have of Balanced Rock is from a website that calls it "a huge boulder that appears precariously balanced."

Hawk Rock is described by some hikers as the "Stonehenge of Putnam County." The rock is a 30-foot-high monolith, and is said to resemble a huge, perched bird, like a hawk.

History

George Baum, chairman of the Kent Conservation Advisory Committee, believes that the carvings of a turtle, long-tailed bird, and the sun on Hawk

Rock date back to the turn of the twentieth century, and may have been created by members of the Hunt family. The carvings are not considered to be the works of early Native Americans.

Others, however, believe differently. Philip J. Imbrogno, a paranormal investigator, contends that the carvings are from antiquity based upon a conversation he had with a local Native American shaman who told him that the symbols had been on the rock for as long as his people could remember. Imbrogno also believes that Hawk Rock possesses spiritual energy that can be felt when visiting the site.

Here is a chance for you to see for yourself.

The trail to Hawk Rock was marked by Patrick LaFontaine as part of his 2015 Eagle Scout project. He also installed the information kiosk. In addition to LaFontaine, other hikers participated as well.

Note: The Mead Farm Trail, also accessible from the trailhead, passes by the foundation of the house and barn ruins of the nineteenth century Moses F. Mead farm.

Directions

From Carmel (junction of Routes 52/Snadbeck Road & 301), drive north on Route 52/Snadbeck Road for 2.0 miles. Turn left onto Route 42/Farmers Mills Road and head west for less than 1.0 mile. Then turn left onto Wangtown Road and proceed southwest for 0.8 mile. Park to your left in a small area near the end of the road.

Follow the Balanced Rock/Hawk Rock Trail south for 1.5 miles. Near the junction of the Balanced Rock/Hawk Rock Trail and the Mead Farm Trail you will encounter the Balanced Rock; then, continuing south for another 0.2 mile, Hawk Rock is reached at 1.7 miles.

Resources

"Hawk Rock." https://kentcac.info/wp/hikes/hawk-rock.

Philip J. Imbrogno, "The Mysteries of Hawk Rock," *Greenwich Time*, September 8, 2010, https://greenwichtime.com/opinion/article/The-mysteries-of-Hawk-Rock-649692.php.

"The Hiking Group – CT & NY (Southbury, CT)" https://meetup.com/The-Hiking-Group-Ct-NY/events/247269648.

Kevin Flynn, "A Hike into the Mystic, or Just a Walk in the Woods?" *New York Times*, October 15, 2010, https://nytimes.com/2010/10/15/nyregion/15hawk.html. The article includes photographs of Hawk Rock.

Brewster High School Boulder 11

Type of Formation: Large Boulder
WOW Factor: 5
Location: Brewster Hill (Putnam County)
Tenth Edition, NYS Atlas & Gazetteer: p. 109, A6
Earlier Edition, NYS Atlas & Gazetteer: p. 33, AB6–7
Parking GPS Coordinates: 41°26.454′N, 73°36.073′W
Brewster High School Boulder GPS Coordinates: 41°26.487′N, 73°36.035′W
Accessibility: 0.05-mile bushwhack/hike
Degree of Difficulty: Moderately easy
Additional Information: No hunting, trapping, fires, motor vehicles, alcohol, tobacco, or firearms are allowed on the property.

Description

The Brewster High School Boulder is a 12–15-foot-high boulder that is part of a smaller contingent of rocks. Some speculate that at one time there was just one mammoth rock before it broke apart into several pieces. The rocks rest at the top of a small ridge overlooking a marshland.

History

The rock is named for its proximity to the Brewster High School. The school was founded in 1971.

Brewster is named after Walter Brewster, a nineteenth-century landowner.

Directions

From Brewster Hill (junction of Routes 62/"Farm to Market Road" & 312), drive north on Route 62/"Farm to Market Road" for 0.5 mile. Turn right onto Foggintown Road and proceed east for 0.3 mile, passing by the Brewster High School (to your left) along the way. Turn left into a small pull-off 0.1 mile past the school.

From the pull-off, walk to the south end of the parking space and head into the woods, following a slope. Within 100 feet, you will come to a posted sign that reads: "You are entering Brewster Central School District. These woods are used for educational and athletic purposes. Use these woods at your own risk." In other words, take responsibility for your own well-being if you enter the property.

Just beyond the posted sign, you will come to an old woods road that is used by the school. Turn right and follow the dirt road counterclockwise for >100 feet to reach the grouping of rocks, on your right.

7. *(opposite)* Brewster High School Boulder.
Photograph by the author.

Laurel Ledges

12

Laurel Ledges Natural Area

Type of Formation: High Ledges; Boulder
WOW Factor: 6
Location: Patterson (Putnam County)
Tenth Edition, NYS Atlas & Gazetteer: p. 109, A6
Earlier Edition, NYS Atlas & Gazetteer: p. 33, A7
Parking GPS Coordinates: 41°29.085′N, 73°36.152′W
Destination GPS Coordinates: *Laurel Ledges:* 41°28.989′N, 73°36.103′W; *Vertical Wall Rock formation:* 41°29.021′N, 73°36.116′W
Accessibility: *Laurel Ledges, along base:* >0.1-mile hike; *Upper section:* >0.2-mile hike/uphill climb, followed by variable mileages depending upon paths chosen
Degree of Difficulty: Moderate

Description

The Laurel Ledges is a 35–40-foot-high cliff face that extends for >0.1 mile. The Putnam County Land Trust website describes it as something unique and dramatic.

The Vertical Wall Rock Formation is part of what makes the Laurel Ledges unique. It stands straight upright, 15 feet high, 12 feet long, and 5 feet wide. It literally looks like a huge slab of rock that has been lowered into place next to the trail.

History

The Laurel Ledges Natural Area, of which Laurel Ledges is part, includes separate parcels for the Sterling Farm, Tom's Path, Turtle Pond, Lushinsky Plunkett, and Peter Harford Dunlop Preserves.

Directions

From Haines Corner (junction of Routes 164 & 22), drive west on Route 164 for 1.7 miles. At Turtle Pond, bear right onto Cornwall Hill Road and proceed north for 0.4 mile, turning into a pull-off on your right opposite Devon Road, where a sign says "Laurel Ledges."

The trail leads off from the parking area and promptly turns right, heading south and then east. You will be constantly walking along and through a talus field at the base of an enormous cliff, sometimes sheered and sometimes shattered.

The Vertical Wall Rock Formation is encountered >0.05 mile from the start, directly to your left.

A great deal of trail work has gone into making this trek easy as opposed to difficult. Stones and rocks stairs have been strategically placed as well as extensive sections of boardwalks where the walk abuts the swampy part of Turtle Pond.

Eventually, you will reach a point where a red-blazed trail enters on your right as it crosses a stream. Continue straight ahead, now on the yellow-marked trail, as it leads up the mountain through a series of switchbacks. Near the top, you will come to a junction with a red-blazed trail. There are many possibilities here. You can follow the red-blazed trail

8. *(opposite)* Natural Rock formation at Laurel Ledges.
Photograph by the author.

to your left, or go right, following the red-blazed trail as it leads to a blue-blazed trail. All the important rocks, however, seem to be below you at this point.

Resources

"Laurel Ledges: Turtle Pond." https://www.adventuresaroundputnam.com/places-to-go/laurel-ledges-turtle-pond.

"Laurel Ledges Natural Area." https://putnamcountylandtrust.org/laurel-ledges-natural-area.

Little Stony Point Mine/Cave

13

Little Stony Point State Park: Hudson Highlands State Park

Type of Formation: Mine
WOW Factor: 6
Location: North of Cold Spring (Putnam County)
Tenth Edition, NYS Atlas & Gazetteer: p. 108, A2
Earlier Edition, NYS Atlas & Gazetteer: p. 32, A4
Parking GPS Coordinates: 41°25.593'N, 73°57.943'W or 41°25.599'N, 73°57.925'W
Little Stony Point Mine/Cave Coordinates: 41°25.491'N, 73°58.113'W (estimated)
Accessibility: 0.4-mile walk (following loop trail counterclockwise)
Hours: Daily, dawn to dusk
Degree of Difficulty: Easy

Description

This mine shaft of unknown name goes horizontally into the rock face of the Little Stony Point cliffs. It is 6 feet wide and starts off at a height of 6 feet but immediately tapers to 4 feet. The cave is enterable (and many do venture in for short distances) but be prepared to duckwalk unless you are a child. Reports indicate that the shaft ends after 75 feet.

Because this opening looks like a cave, many refer to it as the Little Stony Point Cave.

There are very few mines on the east side of the Hudson River. This is not the case on the river's west side, where Bear Mountain Park and Harriman Park list a total of 19 mines on or near its trails. Edward J. Lenik's 1996 book, *Iron Mine Trails*, is a great reference book for those interested in trails that pass by mines on the west side of the Hudson River.

9. Little Stony Point Mine. Photograph by the author.

History

Although little physical evidence remains to indicate otherwise, Little Stony Point started off as an island. During the early 1900s, the channel separating the island from the mainland was filled in by the Hudson River Stone Company as they quarried rock from nearby Bull's Quarry.

Later, the land was acquired by the Georgia Pacific Company, which intended to establish a wallboard factory on the land. Fortunately, New York State intervened, the wallboard factory site was relocated to Verplanck, and in 1970, Little Stony Point opened up to the public as a state park.

Directions

From Cold Spring (junction of Routes 9D/Morris Avenue & 301/Main Street), drive northwest on Route 9D/Morris Avenue for 0.7 mile. Park along the side of the road to your left near the Little Stony Point State Park sign or turn right and park in the large area for the Washburn Trail.

From the kiosk next to the Little Stony Point State Park sign, cross over a bridge that spans the Metro North rail tracks. You will immediately pass by a wide trail on your left. This is your exit point. In <100 feet, you will pass by the Overlook Trail, which, if taken, leads in 0.1 mile to a rocky overlook of the Hudson River. You should take it, for the views of the Hudson River are extraordinary.

Back on the main trail, heading downhill, bear left at a junction and follow the trail as it takes you down to a beach area (where swimming is not allowed). From here, veer left, following along an extensive escarpment wall, first heading southwest and then southeast. The mine shaft is located in the rock wall on the southeast side of Little Stony Point where the trail goes along the base of the cliff wall next to the Hudson River.

From the mine, it is <0.1 mile back to your starting point, on the west side of the bridge spanning the Metro North tracks.

Resources

"Little Stony Point." https://summitpost.org/little-stony-point/357034.

Edward J. Lenik, *Iron Mine Trails (New York: New York–New Jersey Trail Conference, 1996).*

Westchester County

Westchester County is named after the city of Chester, England. The county is bordered by the Hudson River to the west, and by the Long Island Sound and the county of Fairfield, Connecticut, to the east. It encompasses 500 square miles, of which 430 square miles is land, and is the second-most populous county on the mainland of New York State.

Brinton Brooks Sanctuary Rocks 14

Brinton Brook Sanctuary

Type of Formation: Balanced Rock
WOW Factor: Unknown
Location: Croton-on-Hudson (Westchester County)
Tenth Edition, NYS Atlas & Gazetteer: p. 108, D3
Earlier Edition, NYS Atlas & Gazetteer: p. 33, C4–5
Parking GPS Coordinates: 41°13.373'N, 73°54.307'W
Destination GPS Coordinates: *Balanced Rock: Unknown; Split Rock: Unknown*
Accessibility: 0.3–0.4-mile trek (estimated)
Degree of Difficulty: Moderate
Trail Map: sawmillriveraudubon.org/maps/

Description

In *Natural New York*, Bill and Phyllis Thomas write, "Rock ledges with glacial grooves and a boulder balanced atop smaller rocks attract students of geology."

Peggy Turco and Katherine S. Anderson, in *Walks and Rambles* in *Westchester and Fairfield Counties: A Nature Lover's Guide to 36 Parks and Sanctuaries*, write that "the Red Trail leads to one of Brinton's most appealing features—the split rock spring. A stone seat invites you to sit awhile near water bubbling from the split in a huge rock."

History

The Brinton Brook Sanctuary is a 156-acre preserve offering over 4 miles of hiking trails. It is the Saw Mill River Audubon's largest sanctuary.

The sanctuary is named for Willard and Laura Brinton, who donated the first 112 acres to National Audubon to form the nucleus of the wildlife sanctuary. Additional parcels were acquired from the Brintons' niece, Ruth Brinton Perera, in 1975, and by the Saw Mill River Audubon in 1995.

The land is owned by the village of Croton-on-Hudson, and managed by the Saw Mill River Audubon.

The Pond Loop Trail is marked by twenty interpretive signs that focus on the preserve's flora and bird inhabitants.

Directions

Traveling along Route 9 in Croton-on-Hudson, take the Senasqua Road/9A exit. When you come to Route 9A, head north for 1.1 miles (or <0.2 mile northwest of Arrowcrest Drive) When you see the sign for the sanctuary, turn right and drive east for >0.3 mile. Park in the area straight ahead at the point where the road turns sharply left as it heads toward a private residence.

A second entrance [41°13.276'N, 73°54.062'W] is located along Arrowcrest Drive, 0.5 mile from Route 9A, just before the entrance to the Hudson National Golf Club. A sanctuary sign is also present at this trailhead.

Finding the balanced rock mentioned by Bill and Phyllis Thomas is not easy since the sanctuary map makes no reference to a balanced rock, although the Split Rock Spring area is indicated.

To get to the Split Rock Spring area from the parking area, follow the yellow-marked Pond Loop Trail northeast for >0.1 mile. Bear right onto the red-marked Hemlock Spring Trail and head east for 0.05 miles. Finally, turn right onto the yellow-marked Laurel Rock Trail. You will immediately come to Split Rock Spring, where hopefully a geological formation or two await. If not, there are additional trails to explore, but we have no further suggestions to offer.

Resources

Bill and Phyllis Thomas, *Natural New York (New York: Holt, Rinehart and Winston, 1983), 235.*

Peggy Turco and Katherine S. Anderson, Walks and Rambles in Westchester and Fairfield Counties: A Nature Lover's Guide to 36 Parks and Sanctuaries (Woodstock, VT: Backcountry Publications, 1993), 33.

Devil's Stairs 15

Type of Formation: Rock Steps
WOW Factor: Unknown
Location: Ossining (Westchester County)
Tenth Edition, NYS Atlas & Gazetteer: p. 108, D3
Earlier Edition, NYS Atlas & Gazetteer: p. 33, C5
Dale Cemetery Marble Place Entrance GPS Coordinates: 41°10.170′N, 73°51.366′W
Devil's Stairs GPS Coordinates: Not determined
Accessibility: Unknown

Description

In *The Place-Names of Westchester County, New York*, Richard M. Lederer Jr. writes that the Devil's Stairs is "a natural rock formation in Sing Sing Kill southwest of the Marble Place entrance to Dale cemetery."

History

Sing Sing Brook, aka Sint-sinck Brook and Kil Brook, is a small tributary to the Hudson River whose first 6,000 feet, going upstream, are tidal.

The Devil is no stranger to this part of the state, particularly to the Catskills and Hudson Valley where there is the Devil's Path, Devil's Kitchen, Devil's Pulpit, Devil's Hole, Devil's Chamber, and Devil's Tombstone, just to mention a few that come to mind.

Directions

From Ossining (junction of Routes 133/Croton Avenue & 9/Highland Avenue), head northeast on Route 133/Croton Avenue for 0.3 mile. Bear left onto Dale Avenue and continue northeast for 0.2 mile. Then turn left onto Marble Place and proceed northeast until you reach the entrance to 40-acre Dale Cemetery.

The Devil's Stairs are said to be located southwest of the Marble Place entrance to the cemetery, but that's about all we have been able to find out about where to look.

Sing Sing Kill makes a U-turn by the cemetery and then flows through a series of residential areas. The U-turn section of the stream can be observed from the southwest side of the cemetery, but it seems doubtful that the rock formation lies here.

Beyond the cemetery's reach, access is likely to be problematic due to houses and buildings lining the stream. The upper part of Sing Sing Kill is essentially non-navigational, which makes access by kayaking impractical.

However, there is a 0.3-mile-long section of the Sing Sing Kill that is readily accessible and contains some small cascades which possibly could turn into natural steps during the summer. Enter the deep Sing Sing Kill Gorge via the Sing Sing Greenway Trail, a combination of cement walkways, steps, and boardwalks. Start at a parking area off Broadway (41°09.805'N, 73°51.778'W) and end up, 0.3 mile later, at Central Avenue (41°09.632'N, 73°51.976'W).

TO GET TO THE SING SING GREENWAY TRAIL

From Ossining (junction of Route 9, Route 133 & Broadway), drive west on Broadway for 0.1 mile and turn right into the parking area next to the Ossining Aquatic Center.

Resources

Richard M. Lederer Jr., *The Place-Names of Westchester County,* New York (Harrison, NY: Harbor Hill Books, 1978), 40.

Arthur G. Adams, The Hudson River Guidebook (New York: Fordham University Press, 1996), 145.

"Sing Sing Kill Greenway – Ossining Park on the Sing Sing Kill." https://www.inossining.com/parks/sing-sing-kill-greenway.

Devil's Rock and Devil's Footprints

16

Type of Formation: Large Boulder
WOW Factor: Unknown
Location: Croton-on-Hudson (Westchester County)
Tenth Edition, NYS Atlas & Gazetteer: p. 108, D3
Earlier Edition, NYS Atlas & Gazetteer: p. 33, BC4–5
Hessian Hill GPS Coordinates: 41°13.366′N, 73°52.715′W (Google Earth)
Devil's Rock & Devil's Footprints GPS Coordinates: Unknown
Accessibility: Presumably near roadside

FOOTPRINTS IN ROCK PUZZLING BIG SCIENTISTS

Huge Boulder at Croton, N. Y., Bears Imprint of Pair of Human Feet.

WEIRD LEGENDS TOLD

One Attributes Mysterious Footprints to His Satanic Majesty.

CROTON, N. Y., May 2.—Mysterious footprints in the solid rock on the east and west banks of the Hudson here have puzzled the scientists, who be-

10. Headline describing Devil's Rock and Devil's Footprint in the *Syracuse Journal*. *Syracuse Journal*, May 2, 1913, public domain.

Description

The Devil's Rock is a large boulder that bears the imprint of a pair of human feet. In an article that appeared in the *Syracuse Journal in 1913, the author writes,*

> On the east shore, along the old Albany postroad [*sic*] and at the bottom of a steep hill belonging to the A. P. Gardiner estate, lies a huge boulder shadowed by tall trees . . . Its smooth surface bears the imprint of a pair of human feet placed side by side, as if a barefooted man had walked down the hill and stood on the spot while the stone was still soft and yielding from nature's crucible. Every toe is clearly defined, and judging from the mold he left in the granite, the foot of this ancient man was both large and shapely. Behind the footprints, all the way to the top of the rock, are a series of peculiar indentations such as the links of a heavy chain would make in soft earth.

Such a strange rock, it should be pointed out, is not as unique as you might think at first. There are a number of Devil's Rocks and Devil's Footprints in the United States, one of the most famous being the Devil's Rock (with Satan's footprint imbedded in it) in North Carolina.

History

The Devil's Rock has also been referred to as the Devil's Track, presumably because of the footprints.

The rock is what scientists call a *petrosomatoglyph*, a representative part of a human or animal that has formed naturally in rock—in this case the impression of a footprint.

The rock is located on the former Alfred P. Gardiner estate which extended east uphill from the Hudson River for over a mile.

The only other Devil's Rock in New York State that we are familiar with is the Devil's Rock [42°59.306′N, 78°07.408′W] along Route 5 in Batavia where, rumor has it, the Devil was temporarily chained until, trying to escape by madly running around and around the boulder, he finally broke loose, leaving behind a deep groove he had worn into the rock, which is why the boulder is mushroom shaped.

Directions

Even when the article about the Devil's Rock was written in 1906, the exact location of the boulder was already gone from memory.

It's possible that during the last hundred years this fascinating boulder has been rediscovered. If so, we can find no mention of it in the current literature.

To look for the Devil's Rock, we would suggest starting at the base of Hessian Hill along Route 9 which, we believe, is the Old Albany Post Road mentioned in the *Syracuse Journal article. The problem is that there is a considerable amount of terrain to cover, and no specific clues to serve as a guide.*

Resources

Syracuse Journal, May 2, 1913. https:/crotonhistory.org/2013/11/12/the-mystery-of-the-devils-footprints.

"The Mystery of the Devil's Footprint." https:/crotonhistory.org/category/mysteries.

"The Jolly Baker: The Devil's Footprints." https:/thejollybaker.com/2022/02/20/the-devils-footprints.

Potato Rock (Historic) 17

Croton Point Park

Type of Formation: Large Rock
WOW Factor: Unknown
Location: Croton Point (Westchester County)
Tenth Edition, NYS Atlas & Gazetteer: p. 108, D3
Earlier Edition, NYS Atlas & Gazetteer: p. 33, C4–5
Parking GPS Coordinates: 41°11.024′N, 73°53.962′W
Potato Rock GPS Coordinates: 41°11.156′N, 73°54.226′W (Google Earth)
Fee: Modest charge to enter park
Additional Information: Croton Point Park, 1 Croton Point Avenue, Croton-on-Hudson, NY 10520
Croton Point Park Map: parks.westchestergov.com/images/stories/pdfs/Croton.pdf

---

Description

Potato Rock, before its destruction, was a large, oblong-shaped rock located slightly offshore. Its oblong shape and perhaps rough, knobby surface are undoubtedly the features that gave rise to the name Potato Rock.

Google Earth and a couple of websites continue to list the rock's GPS coordinates even though Potato Rock has not existed for some time.

Raymond H. Torrey, Frank Place Jr., and Robert L. Dickinson write in *New York Walk Book*, "Along the shores of both sides of the tip are great quantities of ice- and water- borne boulders deposited in the moraine." On the west side of the tip, facing the Hudson River, are small-to-medium-sized rocks to be seen, but none that will excite the imagination.

History

Potato Rock was blasted to bits sometime in the late 1900s for being a navigational hazard.

Croton Point Park was established about 1924. The 508-acre park is the site of an old Native American village described by James Owen in "The Fortified Indian Village at Croton Point." According to Ann B. Silverman in "Guarding County's Archaeological Past," the site was occupied by Native Americans as far back as 6,300 years ago.

Directions

From Route 9 near Croton Point, take the exit for Croton Point Avenue/Croton Harmon Station. When you come to Croton Point Avenue, drive southwest for 1.2 miles to reach the Croton Point Park parking area.

From the northwest end of the parking area, walk north for >0.1 mile and look out to your left across the Hudson River. There is nothing to see of Potato Rock except to know that you are one of the few people who know both of its existence and where it was located.

Resources

Richard M. Lederer Jr., *The Place-Names of Westchester County, New York* (Harrison, NY: Harbor Hill Books, 1978), 115.

Ann B. Silverman, "Guarding County's Archaeological Past," *New York Times*, November 17, 1985.

Raymond H. Torrey, Frank Place Jr., and Robert L. Dickinson, *New York Walk Book*, 3rd ed. (New York: American Geographical Society, 1951), 74.

James Owen, "The Fortified Indian Village at Crown Point. *The Westchester Historian: Quarterly Bulletin of the Westchester Historian* 2, no. 2 (April 1956): 3.

Frank's Rock 18

Type of Formation: Rock Outcrop
WOW Factor: Unknown
Location: Ossining (Westchester County)
Tenth Edition, NYS Atlas & Gazetteer: p. 108, D3
Earlier Edition, NYS Atlas & Gazetteer: p. 33, C4–5
Parking GPS Coordinates: 41°11.076'N, 73°52.449'W
Frank Rock's GPS Coordinates: 41°11.195'N, 73°52.538'W (estimated)
Accessibility: 0.1-mile walk

Description

There is little specific information about Frank's Rock other than what is listed in the history section. Our suspicion is that Frank's Rock is not a boulder, but rather a rock outcropping. This is based upon a photograph taken of the rock in 1965 by Renoda Hoffman that showed a rock outcrop. Still, it's entirely possible that a rock or small boulder may be associated with the rock outcrop.

In an article entitled "Frank's Rock," Greta Cornell writes, "on the side of the river as it turns towards the lower portion of the bay is a large rock. It has always been a favorite fishing point and is called Frank's Rock."

History

Frank's Rock, aka Frans Besley's Rock, is named for Frans Besley who was born in 1686, and who fished from the rock at the mouth of the Croton River below the Mary Immaculate School.

Directions

From Crotonville (junction of Ogden Road and Old Albany Post Road), drive southwest on Old Albany Post Road for >0.1 mile. As soon as you pass under the Route 9 overpass, turn right at a stop sign onto Eagle Park and proceed >0.2 mile to St. Augustine's Roman Catholic Church and School to park.

Walk north, if access is permitted, to reach the south bank of the Croton River near its confluence with the Hudson River. It's hard to know what you are going to find here. We have provided GPS coordinates to flag a possible candidate for Frank's Rock. However, it's important to bear in mind that nearly 350 years have transpired since Besley fished from the rock. A lot could have happened and probably has since then.

Resources

Richard M. Lederer Jr., *The Place-Names of Westchester County,* New York (Harrison, NY: Harbor Hill Books, 1978), 53.

Greta Cornell, "Frank's Rock," *The Westchester Historian: Quarterly of the Westchester County Historical Society* 41, no. 1 (1965): 11–12. A photograph of the rock taken by Renoda Hoffman can be seen on page 11.

Teatown Lake Boulders 19

Teatown Lake Reservation Nature Preserve

Type of Formation: Large Boulder
WOW Factor: 5–6
Location: Ossining (Westchester County)
Tenth Edition, NYS Atlas & Gazetteer: p. 108, D3–4
Earlier Edition, NYS Atlas & Gazetteer: p. 33, BC5
Parking GPS Coordinates: *Teatown Lake Reservation Nature Preserve Visitor Center:* 41°12.676′N, 73°49.626′W; *Blinn Road:* 41°12.817′N, 73°49.602′W
Tea Town Boulders GPS Coordinates: 41°12.841′N, 73°49.616′W (per rock climbing website); *large rock near Blinn Road parking area: 41°12.816′N, 73°49.607′W. These are just a few of the boulders in the preserve.*
Accessibility: Variable distances
Degree of Difficulty: Easy to moderately easy
Additional Information: Teatown Lake Reservation Nature Preserve, 1600 Spring Valley Road, Ossining, NY 10562
Trail Map: teatown.org/map/

Description

A number of large boulders can be found along the Lake Trail, particularly as you head into the woods from Teatown Lake.

Large boulders are also mentioned near Vernay Lake, which is part of the reservation south of the Visitor Center.

In *Walks and Rambles in Westchester and Fairfield Counties: A Nature Lover's Guide to 36 Parks and Sanctuaries*, Peggy Turco and Katherine S. Anderson write, "In winter spectacular ice formations decorate the many small caves and crannies in the rock," these presumably being talus caves.

11. Teatown Boulders. Photograph by Dan Balogh.

History

The Teatown Lake Reservation Nature Preserve encompasses 1,000 acres of land, including 42-acre Teatown Lake. It contains 15 miles of hiking trails.

The name Teatown dates to 1776 Revolutionary War days when a group of local women who called themselves "Daughters of Eve" boycotted an entrepreneur named John Arthur who wanted to sell his supply of tea at exorbitant prices. They won out, and the area, as a result, became known as Teatown.

The property subsequently passed through several owners, including Arthur Vernay (for whom Vernay Lake is named), who built "The Croft" (a house with a garage that was part of the land purchase), and then later Dan Hanna, whose horse stable eventually became the Nature Center. In 1923, Gerard Swope Sr. purchased the land, created the lake by damming up Bailey Brook, devised a system of trails, and used the land for horseback riding. Six years after Swope's death, his family donated 194 acres of land to the Brooklyn Botanical Garden to establish an outreach station in Ossining. In 1971, Teatown become incorporated, and eventually the tie with the Botanical Garden was broken.

There may also be boulders in nearby Ossining and Sing Sing, although we have not been able to find any references to large rocks through our library research. We mention Ossining and Sing Sing because an online magazine called *Postscript (postscriptmagazine.org/content)* reminds us that the name Ossining is Native American for "a place of stones" (presumably dolomitic limestone outcroppings) and Sing Sing for the "place of rocks."

Directions

VISITOR CENTER PARKING

From Glendale (junction of Spring Valley Road & Glendale Road), drive northeast on Spring Valley Road for 1.4 miles to reach the Teatown Lake Reservation Nature Preserve Visitor Center, on your left.

BLINN ROAD PARKING

From the visitor center, continue southeast on Spring Valley Road for >0.1 mile and turn left onto Blinn Road. Go northwest on Blinn Road for <0.2 mile and turn left onto a dirt road that leads promptly to a parking area.

Many of the rocks can be found along the Lake Trail on the east side of Teatown Lake, and into the woods.

VERNAY LAKE

To reach 9-acre Vernay Lake from the visitor center walk across the road and up to a gate in the stone wall that leads to a kiosk. From there, follow the orange-blazed Twin Lakes Trail south, initially following a gravel road, and then down a stone-step path to the lake. Expect to do some bushwhacking and reconnoitering to find the boulders once you reach the lake. Most are on the opposite side of Vernay Lake.

Resources

Lincoln Diamant, *Teatown Lake Reservation: Images of America* (Charleston, SC: Arcadia Publishing, 2003). The book provides extensive information about the reservation.

"Tea Town Boulders Climbing." https://www.mountainproject.com/area/107497338/tea-town-boulders.

"Tea Town, Westchester County Boulders."www.mountainproject.com/forum/topic/107492270/tea-town-westchester-county-boulders.

"Tea Town Reservation Bouldering." climbingandbouldering.com/new-york/tea town-reservation-bouldering.

Peggy Turco and Katherine S. Anderson, *Walks and Rambles in Westchester and Fairfield Counties: A Nature Lover's Guide to 36 Parks and Sanctuaries* (Woodstock, VT: Backcountry Publications, 1993), 52.

Natural Bridge (Historic) 20

Type of Formation: Natural Bridge
WOW Factor: Unknown
Location: Salem Center (Westchester County)
Tenth Edition, NYS Atlas & Gazetteer: p. 109, BC6
Earlier Edition, NYS Atlas & Gazetteer: p. 33, B6–7
Natural Bridge GPS Coordinates: Unknown

Description

Richard M. Lederer Jr., in *The Place-Names of Westchester County, New York,* makes mention of a natural bridge that once existed near Salem Center but, unfortunately, cites no specifics other than that Titicus Road crossed over it near the intersection of Titicus Road and Delancey Road.

In her article, "A Natural Bridge in Westchester," Allison Albee writes, "Many years ago the bridge was a featured attraction in the area and all traffic between Salem and Purdy's passed over it." When the natural bridge was visited in 1940, it was supposedly almost indistinguishable due to the accumulation of growth. The north side of the bridge was completely filled in, and the south side had a 2-foot-wide opening, with a tiny stream emerging from it. "We are at a loss," wrote Albee, "to reconcile the present approximate twelve-foot height of the rock face with the twenty-five-foot height given in [Robert] Bolton's 1848 description of the spot."

Going back to 1848's *A History of the County of Westchester from Its First Settlement to the Present Time, Robert Bolton Jr. writes,*

> Here are two streams which meet and run under the road, the one flowing from the east along the road side, enters the ground twenty-five or thirty feet east of where it seems to cross the road, the stream from the northeast, appears to run nearly straight, directly under the road and issues from the earth again, after falling ten or fifteen feet

lower than where it enters, but the place where it issues from the earth, is at least twenty-five feet perpendicular, the top of which precipice is within ten or fifteen feet from the side of the road.

It is difficult to imagine what the spot must look like now, over 60 years since Allison Albee's article appeared.

History

The 681-acre Titicus Reservoir, which flooded the valley in 1895 and forced the rerouting of Titicus Road in 1893, obviously changed the topography of the area, as well as causing the natural bridge to fade from memory.

The reservoir is one of twelve in New York City's Croton system.

Directions

From North Salem (junction of Routes 116/Titicus Road & 121/Grant Road), head west on Route 116/Titicus Road for 1.5 miles to reach the intersection of Titicus Road and Delancey Road. This is not the original location of the intersection, which was shifted northward when the reservoir was created. Allison Albee writes that the natural bridge "is located in the Town of North Salem close beside the north shore of Titicus Reservoir."

To be honest, our first assumption was that the natural bridge had been engulfed by waters when the reservoir was created. It was a surprise, then, to learn that although it became part of the watershed, the natural bridge remained above water level.

It is unknown to us whether the natural bridge, or what remains of it, can still be visited. We do see a stream coming down into the reservoir at 41°20.099'N, 73°36.768'W along Titicus Road and can't help but wonder if that might lead to the natural bridge. Whether permission to enter watershed land has to first be obtained could be a factor, however, about proceeding further.

Resources

Richard M. Lederer Jr., *The Place-Names of Westchester County, New York* (Harrison, NY: Harbor Hill Books, 1978), 99.

Allison Albee, "A Natural Bridge in Westchester," *The Westchester Historian of the Westchester County Historical Society* 33, no. 4 (November–December 1957): 114–115.

Robert Bolton Jr., A History of the County of Westchester, from Its First Settlement to the Present Time, vol. 1 (New York: Alexander S. Gould, 1848), 478.

Great Boulder and Old Academy Boulder

21

Type of Formation: Perched Rock; Large boulder
WOW Factor: 8
Location: North Salem (Westchester County)
Tenth Edition, NYS Atlas & Gazetteer: p. 109, BC7
Earlier Edition, NYS Atlas & Gazetteer: p. 33, B7
Destination GPS Coordinates: *Great Boulder*: 41°20.044′N, 73°34.298′W; *Old Academy Boulder*: 41°19.739′N, 73°35.878′W
Accessibility: Roadside
Degree of Difficulty: Easy

12. Great Boulder of North Salem. Antique postcard, public domain.

Description

The Great Boulder, aka Big Rock, Balanced Rock, and the North Salem Dolmen, weighs about 69 tons according to a sign next to the rock, and is supported on five native limestone rocks. Boulders that are supported by smaller, underlying, peg-like rocks are typically called dolmens.

The Great Boulder measures 16 x 14 x 10 feet. Steve Schimmrich, a contemporary geologist, estimates its weight to be 178 tons (almost three times the weight that is indicated on the sign accompanying the rock). Whatever the boulder's actual weight turns out to be, there is no doubt about one thing—this is truly a big rock.

In *History of Westchester County, New York*, Frederic Shonnard and W. W. Spooner write, "It is a prodigious rock of red granite, said to be the solitary one of its kind in the country."

Old Academy Boulder is a medium-sized, 3–4-foot-high boulder that rests in front of the Old Academy at Salem Center. We mention it for its historic value.

History

THE GREAT BOULDER

The Great Boulder is believed to have been carried by glaciers from New Hampshire to its present location and then unceremoniously dropped on top of several small stones.

Others, however, contend that the boulder was propped up by Celts for religious ceremonies long before Christopher Columbus set sail to the New World. Patricia Edwards Clyne in *Hudson Valley Trails and Tales*, seems to support this theory when she writes that "the boulder's seven support stones form an isosceles triangle, the legs of which are in units of a measurement known as the megalithic yard (2.72 feet). The megalithic yard has been cited by various researchers—including Alexander 'Sandy' Thom, who studied structures in the British Isles—as the basic measurement used in laying out dolmens and other Old World megaliths."

On the other hand, Fred C. Warner provides a convincing geological explanation. "With the passing of centuries and the action of the elements, its surrounding and covering material were eroded into the Titicus River and it settled down securely upon the five limestone rocks which just happened to be in a position to support it."

It may be impossible to prove with certainty whether the boulder came to rest naturally in its present position, or whether it was artificially propped up by an ancient tribe centuries ago, but we tend to go with Warner's explanation.

The town of North Salem took possession of the Great Boulder sometime around 1959. The land, originally belonging to Mary A. Quick, was donated to the town by George Cable.

OLD ACADEMY BOULDER

According to the July 1941 issue of the *Quarterly Bulletin of the Westchester County Historical Society*, Charles J. F. Decker "was instrumental in erecting the natural boulder weighing twenty tons and brought one mile and placed in front of the old Academy in memory of the boys of the town who died in the World War."

The historic town hall, formerly the North Salem Academy, was built 1770. From 1790 to 1884 it served as the North Salem Academy; after 1886, it became the town hall.

Directions

GREAT BOULDER

From North Salem (junction of Routes 116 West/Titicus Road & 121/Peach Lake Road), head southwest on Route 116 West/Titicus Road for 0.5 mile (or 0.1 mile past Keeler Lane). The rock is on your left, next to the historic Cable Barn that was erected in 1869.

OLD ACADEMY BOULDER

From the Great Boulder, continue on Route 116 West/Titicus Road for another 0.5 mile. Bear right at a fork (where Route 121/Grant Road goes left) and continue west on Route 116/Titicus Road for another 0.9 mile. Then turn right onto a dead-end road shared by several buildings, including the North Salem Town Police and the Ruth Keeler Memorial Library. The first building on the left is the Old Academy. A boulder, 60 feet south of the building, is visible.

Resources

Patricia Edwards Clyne, *Hudson Valley Trails and Tales* (Woodstock, NY: Overlook Press, 1990), 13–14.

Chris Gethard, *Weird New York: Your Travel Guide to New York's Local Legends and Best Kept Secrets* (New York: Sterling Publishing Co., Inc., 2005), 31.
Frederic Shonnard and W. W. Spooner, *History of Westchester County, New York* (New York: New York History Company, 1900), 15.
Susan Cochran Swanson and Elizabeth Green Fuller, *Westchester County: A Pictorial History* (Norfolk, VA: Donning, 1982) 14, 77.
Frances Eichner and Helen Ferris Tibbets, eds., *When Our Town Was Young: Stories of North Salem's Yesterday* (North Salem, NY: Town of North Salem, 1945). Page 17 includes a photograph of the Great Rock and commentary: "The Boulder, which is estimated to weigh about sixty tons, rests on five smaller stones of limestone and stands about four feet from the surface of the hill on which it is situated."
C. R. Roseberry, *From Niagara to Montauk: The Scenic Pleasures of New York State* (Albany, NY: State University of New York Press, 1982), 334–335.
Westchester Historian: The Quarterly of the Westchester County Historical Society 17, no. 3. (July 1941). A photo of the Great Boulder graces the cover of the bulletin, courtesy of Allison Albee. On page 61 is a photograph of the Academy Boulder, also courtesy of Allison Albee, with further information provided on page 63.
"North Salem's Great Granite Boulder," *Westchester Historian: Quarterly of the Westchester County Historical Society* 3, no. 1 (January–March 1959): 22.
Robert Bolton Jr., *A History of the County of Westchester, from Its First Settlement to the Present Time*, vol. 1 (New York: Alexander S. Gould, 1848). On page 486 is a line drawing of the boulder. Bolton writes, "This immense block viewed from the valley beneath has much the appearance of a huge mammoth ascending the hill."
Fred C. Warner, "North Salem's Great Boulder," *Westchester Historian: Quarterly of the Westchester County Historical Society* 32, no. 1 (January 1956): 29–30. Included with the article is a photograph of the great boulder.
"Mystery of the Giant Boulder in North Salem." https://www.ancientpages.com/2019/08/08/mystery-of-the-giant-boulder-in-north-salem/.

Sarah Bishop's Cave (Historic) 22

Type of Formation: Shelter Cave
WOW Factor: 5
Location: Grant Corner (Westchester County)
Tenth Edition, NYS Atlas & Gazetteer: p. 109, C7
Earlier Edition, NYS Atlas & Gazetteer: p. 33, B7
Sarah Bishop's Cave GPS Coordinates: 41°18.198′N, 73°33.644′W (estimated)
Accessibility: The rock-shelter is on private land
Degree of Difficulty: Unknown
Additional Information: Nearby is Sal J. Prezioso Mountain Lakes Park, 201 Hawley Road, North Salem. In the past, it may have been possible to access the cave from one of the park's trails.

Description

Sarah Bishop's Cave, aka Sarah Bishop's Rock, is a substantial-sized rock-shelter located on a boulder-strewn cliff on West Mountain above the north shore of 35-acre Lake Rippowam.

According to the Discovery Center at Ridgefield, Sarah Bishop's Cave, "was a natural hollow in the rock about 6' square with bark for a door.... Except for a few rags and an old basin, it was unfurnished. [Sarah Bishop's] bed was the floor of the cave and her pillow a projecting point of rock. In a nearby cleft she kept a supply of roots and nuts that she gathered or were given to her by the local townspeople, as she was never a beggar. In the summer, she grew a patch of beans, cucumbers, and potatoes. Nearby were some poor peach trees and numerous highly productive grapevines."

Some of this description sounds questionable. For instance, who could possibly make due with a projecting point of rock for a pillow night after night?

In the January–April 1943 issue of the *Westchester Historian: Quarterly of the Westchester County Historical Society*, the anonymous author states

13. Sarah Bishop's Cave. *Source:* Robert Bolton Jr., *A History of the County of Westchester, from its First Settlement to the Present Time*, vol. 1 (New York: Alexander S. Gould, 1848), 278. Public Domain.

that "the cave was but four feet high and three feet wide, only large enough for one person." A huge rock extended over the entrance that served as protection from the elements.

History

Sarah Bishop was a real life hermitess who lived during the Revolutionary War days and is said to have spent thirty years living in the cave. No one knows for sure why she withdrew from the world at large, but it seems likely that she must have experienced some kind of emotional or physical trauma when she was a young woman. In the same issue of the *Westchester Historian* Sarah Bishop is described as a "young woman, slender and of medium height, fair complexion and graceful figure." Bishop's appearance changed significantly over the years and she died about 1810 from either sickness or exposure. She was buried at the June Road Cemetery [41°19.825′N, 73°35.401′W] in North Salem, where her gravesite is marked by a small plaque.

Richard M. Lederer Jr., in *The Place-Names of Westchester County, New York*, states that Sarah Bishop's Cave "was known in a 1703 deed as

'The Cave.' Archaeologists, excavating for Indian artifacts, found nothing but bacon grease."

From accounts, the cave was somewhat of a tourist attraction during the nineteenth century.

There was enough interest in the cave and Sarah Bishop as a person of curiosity for Fred C. Warner to write a poem that began, "In the forest deep, a rocky cell / was carved by Nature long ago / A lonely spot wherein would dwell / A maid beset by grief and woe."

Directions

From east of Salem Center (junction of Routes 121 & 116 West), drive southwest on Route 121 for 0.8 mile until you come to Grant Corner. Turn left onto Hawley Road and proceed southeast for 1.5 miles. Then turn left onto Mountain Lakes Camp Road. This is as far as we can take you.

We have not been able to find any specific directions on how to get to the cave, which is not surprising since it appears to be located on private property. When Patricia Edwards Clyne wrote *Caves for Kids in Historic New York* 40 years ago, it was possible to access the cave with permission from the Mountain Lakes Camp (as the Sal J. Prezioso Mountain Lakes Park was known back then), but a lot has changed in 40 years.

One website indicates that the Westchester County Parks Commission occasionally leads hikes to the cave, with the landowner's permission, of course. This might be one possible future way of visiting Sarah Bishop's Cave.

Resources

Maureen Koehl, *Lewisboro: Images of America* (Charleston, SC: Arcadia Publishing, 1997). On page 72 is a photograph of the cave taken in 1890.

Patricia Edwards Clyne, *Caves for Kids in Historic New York* (Monroe, NY: Library Research Associates, 1980), 53–54. A photograph of the cave is shown on page 46.

"Sarah Bishop's Cave." https://manuscripts.wordpress.com/2013/11/26/sarah-bishops-cave/.

"Women of the Streets: Sarah Bishop." https://ridgefieldhistoricalsociety.org/ridgefield-women-to-be-remembered-sarah-bishop/.

Fred C. Warner, "Lady of the Cave." *Westchester Historian of the Westchester County Historical Society* 40, no. 3 (1964). This issue contains the full poem that Warner had written about Sarah Bishop.

"The Hermitess of Salem." *The Westchester Historian: Quarterly Bulletin of the Westchester County Historical Society* 19, no. 1 & 2 (January–April 1943): 14–18.

Richard M. Lederer Jr., *The Place-Names of Westchester County, New York* (Harrison, NY: Harbor Hill Books, 1978), 126.

Split Rock 23

Blue Mountain Reservation

Type of Formation: Split Rock
WOW Factor: 4
Location: Peekskill (Westchester County)
Tenth Edition, NYS Atlas & Gazetteer: p. 108, C2–3
Earlier Edition, NYS Atlas & Gazetteer: p. 33, BC4–5
Parking GPS Coordinates: 41°16.199′N, 73°55.305′W
Split Rock GPS Coordinates: 41°15.890′N, 73°54.999′W (estimated)
Fee: Modest admission fee charged
Accessibility: <0.5-mile hike
Degree of Difficulty: Moderate
Additional Information: Blue Mountain Reservation, Welcher Avenue, Peekskill, NY 10566
Trail Maps: parks.westchestergov.com/ images/stories/pdfs/BlueMtn_map_2015.pdf / leathermansloop.org/download/trailmap/BlueMountain05.pdf

---

Description

There are a number of large rocks at this site. The biggest one is <20 feet high, with a 2-foot-long overhang facing the trail. Next to it, on the slope heading down to the lake, is an 8-foot-high rock that is somewhat spherical in shape. Just beyond that, near the lake, is a 6-foot, irregular-shaped rock.

Next to the trail, in front of the main boulder, is an 8-foot-high chunk of rock that has evidently broken off from the main boulder. We suppose that some could call this a split rock, even though the two pieces are separated by a good 10 feet. The *Day Walker: 32 Hikes in the New York Metropolitan Area* apparently thought so, which called it "a very large split rock."

In *Walks and Rambles in Westchester and Fairfield Counties: A Nature Lover's Guide to 36 Parks and Sanctuaries*, Peggy Turco and Katherine S.

14. Blue Mountain Reservation Boulder. Photograph by the author.

Anderson write, "Blue Mountain Reservation is noted for its magnificent rock formations of Hudson Highland granite."

History

The original 1,538 acres of the Blue Mountain Reservation were once farmlands owned by several families. Later, to take advantage of the growing need for keeping produce refrigerated, two artificial ponds—Loundsbury Pond and New Pond—were created by the Loundsbury family, and ice harvesting became a bustling business. With the advent of modern refrigeration, however, the ice-harvesting industry waned, and the property, having lost its value, was acquired by the county in 1926.

The reservation has become a popular mecca for mountain bikers due to its elaborate trail system.

Directions

From Peekskill (junction of Routes 9/Croton Expressway & 6), drive south on Route 9/Croton Expressway for 1.4 miles and take the exit for Welcher Avenue/Buchanan/Route 9A. When you come to Welcher Avenue at a traffic intersection, turn left, go under the Route 9/Croton Expressway, and drive east for 0.4 mile to reach the park entrance. From the contact station, bear left and continue on the park road for another 0.5 mile. Park near the kiosk and trailhead before reaching the end of the road.

From the kiosk, follow the blue/white-marked Blue Mountain Summit Trail southeast uphill, immediately passing by the yellow-marked Dickey Brook Trail, to your left, and the orange-marked Hip Hop Trail to your right. When you come to the junction indicated by post #5 after >0.2 mile, turn right onto the white-marked trail as the blue-marked Blue Mountain Summit Trail veers left.

The white-marked trail takes you downhill and, after a moment or two, passes by the southwest corner of New Pond. Several hundred feet after you cross over the pond's tiny outlet stream, look for a grouping of large rocks on your left. They are impossible to miss. If you come to post #13 by a junction, then you have gone slightly too far.

Resources

New York-New Jersey Trail Conference, *Day Walker: 32 Hikes in the New York Metropolitan Area*, 2nd ed. (Mahwah, NJ: New York-New Jersey Trail Conference, 2002), 147

"Blue Mountain Reservation." https://parks.westchestergov.com/blue-mountain-reservation.

Peggy Turco and Katherine S. Anderson, *Walks and Rambles in Westchester and Fairfield Counties: A Nature Lover's Guide to 36 Parks and Sanctuaries* (Woodstock, VT: Backcountry Publications, 1993), 20.

Hunter Brook Rock Shelter 24

Type of Formation: Rock-Shelter
WOW Factor: 4
Location: Croton Reservoir (Westchester County)
Tenth Edition, NYS Atlas & Gazetteer: p. 108, C3
Earlier Edition, NYS Atlas & Gazetteer: p. 33, BC5
Parking GPS Coordinates: 41°15.585′N, 73°50.468′W (alongside of road)
Destination GPS Coordinates: *Hunter Brook Rock Shelter*: 41°15.584′N, 73°50.499′W (estimated); *Second site*: 41°15.437′N, 73°49.812′W
Accessibility: Near roadside; may be on private land
Degree of Difficulty: Unknown, but probably easy

Description

According to Roberta Wingerson (a professional archaeologist), in her article "The Hunter Brook Rockshelter," "The shelter was formed by the overlap of several large rock slabs probably torn from the ridge above as the glacier moved across it in a southerly direction." Restored through excavation, it is presently 6–8 feet wide and 10 feet long, with adequate headroom for occupants. It would seem that the rock shelter is located at the extreme northwestern end of the 2,182-acre Croton Reservoir.

Due to the creation of the New Croton Reservoir in 1904–1905, waters flowing through the Hunter Brook Valley have been backed up to within 50 feet of the rock-shelter. The immediate surroundings are now much different than they were for early Native Americans.

On the other hand, Caroline (last name not given), in her website https//ossininghistoryontherun.com/2017/06/07/the-hunterbrook-rock-shelter, has come up with another possible site for the rock-shelter, which she observed while running on Baptist Church Road. A photograph on her website shows a clumping of large rocks. While it's possible that this may be the same rock shelter written up by Roberta Wingerson, only at a different location, it's much likelier that it is a second rock shelter, coincidentally along the same road.

History

Archaeological digs have revealed that the Hunter Brook Rock-Shelter was used by bands of Native American hunters or small families moving through the valley.

Directions

From the Taconic State Parkway west of Yorktown Heights, take the Underhill Avenue Exit. Drive southwest on Underhill Avenue for 0.2–0.4 mile (depending upon which direction you exited). Turn right onto Baldwin Road and head north for 0.3 mile.

Then turn left onto Baptist Church Road and drive west for 1.8 miles. If the GPS coordinates that we borrowed from Roberta Wingerson's article are correct [adjustments had to be made], then the rock-shelter will be directly to your right, just before crossing over a short bridge that spans Hunter Brook.

SECOND SITE

The rock-shelter described on https://ossininghistoryontherun.com/2017/06/07/the-hunterbrook-rock-shelter is located on a different section of Baptist Church Road. Based upon a Google Map made by Caroline, we place the location of the rock-shelter at 41°15.437'N, 73°49.812'W.

To access this site, drive east on Baptist Church Road from the bridge at the first site for 0.6–0.7 mile. Look for the rock-shelter to your left.

Bear in mind that both sites may be on posted land. If so, then go no further.

Resources

Roberta Wingerson, "The Hunter Brook Rockshelter" *The Bulletin: New York Archaeological Association*, no. 68 (November 1976): 19–26.

"The Hunter Brook Rockshelter – Ossining History on the Run." https://ossininghistoryontherun.com/2017/06/07/the-hunterbrook-rock-shelter.

Giant Boulder

25

Granite Knolls Park

Type of Formation: Large Boulder
WOW Factor: 6
Location: Yorktown (Westchester County)
Tenth Edition, NYS Atlas & Gazetteer: p. 108, C4
Earlier Edition, NYS Atlas & Gazetteer: p. 33, B5
Parking GPS Coordinates: 41°18.571'N, 73°49.249'W
Destination GPS Coordinates: *Giant Boulder*: 41°18.693'N, 73°49.826'W;
Second rock site: 41°18.681'N, 73°49.754'W
Accessibility: 1.0-mile hike
Hours: Dawn to dusk
Degree of Difficulty: Moderate
Trail Maps: leathermansloop.org/download/trailmap/granite-knolls-park
-trail-map.pdf
nynjtc.org/map/granite-knolls-park

Description

Giant Boulder is a large, near house-sized glacial erratic that lies within a small-scale quarry. The front end of the boulder is 15 feet high and virtually circular in shape. The rock looks to be as long as a whale, extending for a length of 50 feet. Makeshift ladders can be seen propped up against the boulder to assist those who wish to climb to the top.

A number of other large boulders also populate the area.

History

Granite Knolls Park was purchased by the town of Yorktown in 2010.

The area was originally owned by the Jesuits, who farmed the land. The property was also actively quarried. It's said that some of the quarrymen

15. Giant Boulder of Granite Knolls Park. Photograph by the author.

etched their names into the top of Giant Boulder. This would definitely be worth checking out when visiting the rock.

Directions

From the Taconic State Parkway, take Exit 17A and drive east on Route 202 for 0.7 mile. Turn left onto Route 132/Old Yorktown Road and proceed north for 1.5 miles. Then turn left onto Strang Boulevard (on some maps listed as Hunters Brook), head west for 0.5 mile, and park at Woodlands Legacy Field Park.

From Peekskill (junction of Routes 202/35 & 9), drive east on Route 202 for 6.8 miles. Turn left onto Yorktown Road and follow the directions given above.

From the Woodlands Legacy Field kiosk, walk downhill along a gas line pipe corridor. Near the bottom of the hill, bear right onto a blue-marked path that takes you into the woods and then quickly across a stone footbridge constructed in 2007 that spans the Taconic State Parkway.

Turn right at the end of the footbridge, following a wide path north as it parallels the Taconic State Parkway. Within 150 feet, bear left onto the pink-marked Taconic Bridge Trail and follow it west as it completes a series of switchbacks to get you to higher terrain. When you come to the Purple-blazed Dynamite Run Trail, turn left. Within a few moments, the white-blazed Circolara Trail comes in on your right. Stay left on the white-blazed Circolara Trail. Note: *circolara* is the Italian word for circular.

For the next 0.5+ mile, follow the white-blazed Circolara Trail as it takes you in a westerly direction, passing by several junctions. Eventually, the purple-blazed Giant Boulder Trail enters on your left and joins with the Circolara Trail. Continue on the Circolara/Giant Boulder Trail, heading north.

At a junction, bear right onto the purple-blazed Giant Boulder Trail as it separates from the white-blazed Circolara Trail.

Within 0.1 mile you will come to the quarry containing the Giant Boulder.

Resources

"Granite Knolls Park East." https://www.yorktownny.org/community/granite-knolls-park-east.

"Erratic Behavior at Granite Knolls." https://www.geocaching.com/geocache/GC2XQBM_erratic-behaviour-at-granite-knolls.

"Granite Knolls Easy Loop." https://www.nynjtc.org/hike/granite-knolls-easy-loop.

Katonah Woods Boulders and Rocking Stone

26

Type of Formation: Large Boulder
WOW Factor: 4
Location: Katonah (Westchester County)
Tenth Edition, NYS Atlas & Gazetteer: p. 108, C5
Earlier Edition, NYS Atlas & Gazetteer: p. 33, BC6
Destination GPS Coordinates: *Katonah Woods Boulders*: 41°14.627′N, 73°39.527′W (estimated); *Rocking Stone*: Unknown
Accessibility: 200-foot trek (estimated)
Degree of Difficulty: Moderately easy
Additional information: Access may be restricted

Description

The Katonah Woods Boulders are two large boulders that supposedly mark the burial site of Chief Katonah and his wife, Cantitoe.

Of the Rocking Stone, Frances R. Duncombe et al. in *Katonah: The History of a New York Village and Its People* write, "In the woods north of Old Katonah and of great interest to children of the late eighteen hundreds was a 'rocking stone' which they believed had been used by Indians to call one another. According to Margery Van Tassel, it 'could be made to teeter back and forth and make a deliciously thrilling thud of a noise as it hit the underlying rock bed.'"

History

Katonah is named for Chief Katonah, a Native American sachem. Katonah's wife was occasionally called Cantitoe; then at other times, Mustato, according to an online magazine called *Postscript* (postscriptmagazine.org). Katonah is a shortened form of *Ketatonah*, meaning "great mountain."

In the beginning, we assumed that Katonah Woods and its boulders may no longer be accessible. After all, in 1897 the valley was flooded to create the Cross River Reservoir, and the original 50-building hamlet of Katonah (called Old Katonah) was relocated to its present site. By good fortune, however, we came across an article written years after the valley was flooded that gave directions on how to reach the rocks.

Directions

KATONAH WOODS BOULDERS

From I-686/Saw Mill Parkway at Katonah, take Exit 6 for Route 35/Katona/Cross River and head east on Route 35/Cross River Road for 0.4 mile. Turn right onto Route 22/Golden Bridge Road/Jay Street and head southeast for 1.5 miles. Then turn right onto Katonah Woods Road, just before reaching the John Jay Homestead Historic Site (which is also on the right).

From here, we must turn to Arthur I. Bernhard's "Katonah and Bedford: A Do-It-Yourself Historical Tour" for guidance on what to do next. Start by following Katonah Road south. "Where the road turns left downgrade. Stop. Park and walk the narrow branch at right a short distance. On left side—two large rocks—Legend says Indian Chief Katonah and his wife are buried here."

Our interpretation of Bernhard's directions is to drive south on Katonah Road for <0.4 mile and use that as your starting point (41°14.666'N, 73°39.517'W). From here, Bernhard's directions sound obtuse but may make total sense when you are at the spot.

Should the land next to the road be posted, be respectful of private property and go no further.

ROCKING STONE

We have not been able to research the location of this perched rock, which probably has been destabilized over the years by playful youths in a way similar to many small glacial erratics that were pushed over the edge of the great Wall of Manitou in the Catskills.

Resources

Frances R. Duncombe and the Historical Committee, Katonah Village Improvement Society, *Katonah: The History of a New York Village and Its People* (Katonah, NY: The Society, 1961). On page 15 is a photograph of two medium-to-large-size boulders in the Katonah Woods that supposedly mark the graves of Chief Katonah and his wife. On page 7, reference is made to the "rocking stone."

Wikipedia, s.v. "Katonah, New York," last modified January 21, 2023, https://en.wikipedia.org/wiki/Katonah,_New_York.

Arthur I. Bernhard, "Katonah and Bedford: A Do-It-Yourself Historical Tour." *The Westchester Historian: Quarterly of the Westchester County Historical Society* 43, no. 2 (Spring 1967): 32. The cover of this issue shows a photograph of the Katonah Woods boulders that was taken by Renoda Hoffman.

Postscript Magazine. https://postscriptmagazine.org.

Cobbling Rock (Historic) 27

Type of Formation: Large perched rock
WOW Factor: 7
Location: Northwest of Katona (Westchester County)
Tenth Edition, NYS Atlas & Gazetteer: p. 108, C5
Earlier Edition, NYS Atlas & Gazetteer: p. 33, B5–6
Parking GPS Coordinates for Lasdon Park, Arboretum & Veterans Memorial: 41°16.589′N, 73°44.253′W
Cobbling Rock GPS Coordinates: 41.274404°, -73.729240°, or 41.274405, -73.729228.
Accessibility: The rock is located on private property
Additional Information: Lasdon Park, 2610 Route 35, Katonah, NY 10536; (914) 864-7263; Friends of Lasdon Park and Arboretum (FLPA): lasdonpark.org

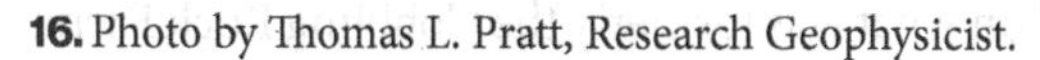

16. Photo by Thomas L. Pratt, Research Geophysicist.

Description

Cobbling Rock, aka Cobbling Stone, is described in J. Thomas Scharf's *History of Westchester County, New York*, as

> a remarkable boulder . . . found in Somers. It stands on the hill directly northeast of Muscoot Mountain in the southwestern part of Somers, and from its top can be seen the blue hills of Long Island across the sound . . . One side of this curious rock has the appearance of an Indian face. It is an immense mass of red granite, said to be the only specimen in the county, and is perched upon three lime-stone points, two or more feet above the surface of the ground.

Sounds a bit like the Great Boulder in North Salem, doesn't it? Except for the location, which doesn't match at all.

The Somers Historical Society, in *Somers: Its People and Places 1788–1988, a 200 Year History*, writes that the rock is "hollowed on one end and it is reputed to have been an Indian grinding mill."

The authors also state that "another reported Indian mortar is located on property owned by Ella Barlow Siemerling on the west side of Mahopic Avenue." (GPS of general area: 41°18.232′N, 73°45.595′W).

History

Cobbling Rock is located on what was formerly the Cobbling Rock Farm, property that was obviously named for its distinctive perched rock. We don't know much about the farmland's early history, but in 1933 a rhinologist and otolaryngologist named Dr. Antone P. Voislawsky erected a main house on the property, replacing one that had been bult in 1903 and later destroyed by fire.

In 1939, the property was acquired by William Lasdon (who made his money through pharmaceuticals) and his wife, Mildred. They used the farm as a country retreat.

When William Lasdon died in 1986, Westchester County purchased the estate, calling it Lasdon Park. At that time, Cobbling Rock was still part of the 234-acre park. According to Gigi Carnes, curator of Lasdon Park, Arboretum, and Veterans Memorial, the land encompassing Cobbling Rock was sold to a private landowner some time ago and is no longer part of the park.

Still, there are many good reasons to visit the park today, which includes the 20-acre Mildred D. Lasdon Bird and Nature Sanctuary, a Chinese

Friendship Pavilion, and the 0.1-acre William and Mildred Lasdon Memorial Garden, established by the Lasdons' daughter, Mrs. Nanette Laitman.

If you can locate Cobbling Rock's private landowner and get permission to cross his land, you may be able to visit the historic rock near Lasdon Park.

Directions

From I-684 in Katona, take Exit 6 for Route 35/Katona/Cross River and proceed northwest on Route 35/Amawalk Road for 3.4 miles. At the sign for Lasdon Park Arboretum and Veterans Memorial, turn left onto Orchard Hill Road and proceed south for <0.4 mile to the parking area.

From Amawalk (junction of Route 35/Amawalk Road & U.S. 202), drive southeast on Route 35/Amawalk Road for 1.5 miles and turn right at the park entrance onto Orchard Hill Road. Head south for <0.4 mile to the parking area.

Bear in mind that the rock is on private property adjacent to the park and is not accessible to the public. Still, there is aways the possibility that the private landowner may grant individuals permission to visit the rock if asked.

Resources

John Thomas Scharf, *History of Westchester County*, vol. 2 (New York: L. E. Preston, 1886), 499. The author describes the rock as a "genuine natural curiosity."

John Thomas Scharf, *History of Westchester County, New York*, vol. 1 (Philadelphia: L. E. Preston, 1886), 9. A photograph of the Cobbling-Rock is shown on page 8.

Somers Historical Society, *Somers: Its People and Places 1788–1988, a 200 Year History* (Somers, NY: Somers Historical Society, 1989). On page 48, the authors write, "On a hill northeast of Muscoot Mountain in southwestern Somers on land owned by the New York City Water District lies Cobbling Rock, an immense mass of red granite."

Great Stone Face Rock Shelter 28

Type of Formation: Rock-Shelter
WOW Factor: 6
Location: Bedford (Westchester County)
Tenth Edition, NYS Atlas & Gazetteer: p. 109, D6
Earlier Edition, NYS Atlas & Gazetteer: p. 33, C6
Bedford Center Coordinates: 41°12.234′N, 73°38.625′W
Tiny lake southwest of Bedford Center GPS Coordinates: 41°11.411′N, 73°39.201′W
Accessibility: Unknown
Degree of Difficulty: Unknown

Description

According to the *Quarterly Bulletin: Westchester County Historical Society*, the Great Stone Face Rock Shelter is a "massive gargoyle like head which protruded above the shelter and formed a part of it and which fancy could liken to a frowning stone giant."

History

Native American artifacts have been found at the site, which makes it historically significant.

Directions

The article on the Great Stone Face Rock Shelter mentions that the geological formation is located 1.25 miles southwest of the old courthouse in Bedford. The old courthouse, built in 1787, now occupied by the Bedford Museum and Bedford Historical Society (615 Old Post Road), is at a GPS of 41°12.234′N, 73°38.625′W. If we draw a straight line southwest for 1.2 miles from the Bedford Museum, we end up near the south end of a small, unnamed lake. Hopefully, this primitive exercise in geometry has put us in the general area of the Great Stone Face Rock Shelter.

Just west of the lake, 0.1 mile distant, is a small quarry; to the east, within 0.1 mile, are residential homes. Having noted this, the best way to explore this area, if possible, is to head uphill from the southwest side of the lake. The elevation here changes rapidly from 343 feet to 422 feet—a 99-foot gain in just 120 feet, sufficient height to produce a rock-shelter. We make no guarantees, but we can't help but wonder just what might be here to find.

To reach the small lake from Bedford Center (junction of Old Post Road & Pound Ridge Road), drive southwest on Old Post Road for 1.1 miles. Turn left onto Crusher Road and go less than 0.3 mile. Perhaps, there is an off-road pull-off before the end of the road, or even at the Town of Bedford Highway Department Facility, from where you can hike east for 0.1 mile, over to the small lake, and then have an opportunity to check out what looks to be a very steep slope (assuming, of course, that the land isn't posted).

Resources

"Great Stone Face Rock Shelter," *Quarterly Bulletin of the Westchester County Historical Society* 7, no. 1 (January 1931). On page 39 is a photo of the rock-shelter taken by Leslie Verne Case. The caption reads "The 'Great Stone Face Rock Shelter' Bedford, NY."

Ward Pound Ridge Reservation Rocks

29

Ward Pound Ridge Reservation

Type of Formation: Rock-Shelter; Boulder, Petroglyph
WOW Factor: *Leatherman's Cave*: 5, *Bear Rock Petroglyph*: 3, *Indian Rock-shelter*: 3–4
Location: Cross River (Westchester County)
Tenth Edition, NYS Atlas & Gazetteer: p. 109, D6
Earlier Edition, NYS Atlas & Gazetteer: p. 33, BC7
Parking GPS Coordinates: 41°14.886′N, 73°35.689′W
Destination GPS Coordinates: *Leatherman's Cave*: 41°14.518′N, 73°36.172′W; *Indian Rock-Shelter*: 41°14.632′N, 73°34.586′W (estimated); *Bear Rock Petroglyph*: 41°13.735′N, 73°35.591′W (estimated)
Fee: Modest entry fee
Accessibility: *Leatherman Cave*: 1.3-mile hike; *Indian Rock-Shelter*: 1.5-mile hike; *Bear Rock Petroglyph*: 1.4-mile hike
Additional Information: Ward Pound Ridge Reservation, Routes 35 & 121 South, Cross River, NY 10518
Park Maps: greatruns.com/wp-content/uploads/2021/01/Ward-Pound-Reservation.pdf
parks.westchestergov.com/images/stories/pdfs/WPRsm_2012.pdf

Description

Leatherman Cave lies on the south shoulder of Overlook Hill, not far below its summit. The cave consists of a jumble of boulders that have fallen together to form a small enclosure.

This is one of the shelter caves reputedly used by the legendary Leatherman during the late 1800s.

17. Bear Rock and its petroglyph. Photograph by Daniel Chazin.

Indian Rock-Shelter consists of an overhanging ledge that archaeologists have determined was used by Native Americans as a hunting camp.

Bear Rock Petroglyph is a gumdrop-shaped boulder that contains a series of squiggles carved into its rock face. The squiggles, which are believed to date back many hundreds of years, depict the shape of a bear, turkey, and deer—all of which were plentiful at the time Native Americans hunted the woods. Interestingly, the petroglyphs were not "discovered" until the park's curator came upon them in 1971.

In *50 Hikes in the Lower Hudson Valley*, Stella Green and H. Neil Zimmerman sound a cautionary note about the petroglyphs. Jim Swager, author of *Petroglyphs in America*, examined Bear Rock and found the authenticity of the petroglyphs questionable.

Additional Boulders: Near the Michigan Road entrance are two large rocks that Peggy Turco and Katherine S. Anderson write about in their book *Walks and Rambles in Westchester and Fairfield Counties: A Nature Lover's Guide to 36 Parks and Sanctuaries*. "The big boulders on top [of either a drumlin or a recessional moraine], to the right of the shelter, are glacial erratics, likewise dumped."

There are also rock destinations called Castle Rock (a rock outcropping), Magie Stairs, Spy Rock, and Raven Rocks, but we don't have any specific information on these geological features other than that below Raven Rocks is a cavern called the Devil's Den.

A Dancing Rock is also contained in the preserve. According to Richard M. Lederer Jr., in *The Place-Names of Westchester County, New York*, "To keep harvest workers warm on cold autumn evenings, a fire was built on a large flat rock. When the work was done there was dancing on the rock."

History

The Ward Pound Ridge Reservation encompasses 4,315 acres of land and is the largest park in Westchester County. Raymond H. Torrey, Frank Place Jr. and Robert L. Dickinson, in *New York Walk Book*, describe it as elliptically shaped, 3.5 miles by 4.5 miles with a circumference of 15 miles. It was first occupied by Native Americans who used "pounds" (or enclosures) to confine captured game. It was undoubtedly this word that later gave rise to the area being called Poundridge.

In 1938, "Ward" was added on to Poundridge to honor a twentieth-century Westchester County Republican leader named William Lukens Ward.

The part of the park that is of most interest to us is found in the southern section.

Directions

From Cross River (junction of Routes 121/Cross River Road & 35), drive south on Route 121/Cross River Road for 0.1 mile, crossing over the Cross River. Turn left onto Reservation Road and proceed east for 0.7 mile until you reach the park's entrance. Once passing through, continue east for another 0.1 mile. Then turn right onto Michigan Road and follow it south for over 0.8 mile to a parking area just before a cul-de-sac.

Near the kiosk is a colored map of the area.

LEATHERMAN'S CAVE

From the Michigan Road parking area, head south and then follow the red-blazed/green-blazed trail southwest past posts 70, 53, and 54. When you come to the Leatherman Trail (LT) at post 31, turn right and head northwest to post 26. Then turn left, proceed downhill south (avoiding going over the summit of 665-foot-high Overlook Hill), and then turn right at post 28, now heading west again. Quickly, you will come to the spur path on your right that leads quickly up to Leatherman's Cave.

The park's map also shows a much shorter path to the cave from Honey Hollow Road. Unfortunately, parking along Honey Hollow Road is extremely difficult unless you park in the Richards Preserve and backtrack north on Honey Hollow Road for 0.2 mile to the park's west entrance.

INDIAN ROCK-SHELTER

From the Michigan Road parking area, head south. At post 70, turn left onto the yellow-blazed/red-blazed trail, heading east. At post 24, where the two trails separate, bear right and follow the yellow-blazed trail south. At post 22, turn left, staying on the yellow-blazed trail (which is now also the Rock Trail). Go past posts 33 and 20. When you come to post 19 as the yellow-blazed trail goes left and heads north, turn right onto the Rock Trail (RT) and head southeast. You will come to the Indian Rock Shelter, on your left, before post 18 is reached.

BEAR ROCK PETROGLYPH

From the Michigan Road parking area, head south. At post 70, proceed south on the green-blazed/red-blazed trail. At post 53, where the two trails separate, bear left and follow the green-blazed trail south. When you come to post 35, stay right, continuing on the green-blazed/red-blazed trail.

At post 38, go left as the green-blazed/red-blazed trail heads north. When you come to post 39, turn left onto the Rock Trail (RT) and head east. You will quickly come to the Bear Rock Petroglyph on your left, not far from a power line corridor.

DANCING ROCK

From the Bear Rock Petroglyphs continue northeast on the Rock Trail (RT). At post 60, turn right and head southeast, going steadily uphill, to reach Dancing Rock at an elevation of 837 feet.

Resources

New York-New Jersey Trail Conference, *Day Walker: 32 Hikes in the New York Metropolitan Area,* 2nd ed. (Mahwah, NJ: New York-New Jersey Trail Conference, 2002), 133.

New York-New Jersey Trail Conference, *New York Walk Book*, 6th ed. (New York: New York-New Jersey Trail Conference, 1998), 111.

Patricia Edwards Clyne, *Caves for Kids in Historic New York* (Monroe, NY: Library Research Associates, 1980), 29.
Peggy Turco and Katherine S. Anderson, *Walks and Rambles in Westchester and Fairfield Counties: A Nature Lover's Guide to 36 Parks and Sanctuaries* (Woodstock, VT: Backcountry Publications, 1993), 108.
Richard M. Lederer Jr., *The Place-Names of Westchester County, New York* (Harrison, NY: Harbor Hill Books, 1978). Dancing Rock is mentioned on page 37.
Patricia Edwards Clyne, *Hudson Valley Trails and Tales* (Woodstock, NY: Overlook Press, 1990), 34.
Beth Herr and Maureen Koehl, *Ward Pound Ridge Reservation: Images of America* (Charleston, SC: Arcadia Press, 213). The entire book is devoted to images of the park.
Stella Green and H. Neil Zimmerman, *50 Hikes in the Lower Hudson Valley* (Woodstock, VT: Backcountry Guides, 2002), 39. The story of the Leatherman is described in some detail. On page 34 is a rather striking photograph of Leatherman's Cave, complete with a hiker sitting on a rock at the entrance.
Daniel Case, *AMC's Best Day Hikes near New York City* (Boston: Appalachian Mountain Club, 2010), 84.
Raymond H. Torrey, Frank Place Jr. and Robert L. Dickinson, *New York Walk Book*. 3rd ed. (New York: American Geographical Society, 1951), 76–79.
"Ward Pound Ridge Reservation." https://parks.westchestergov.com/ward-pound-ridge-reservation.

Henry Morgenthau Preserve Boulder

30

Henry Morgenthau Preserve

Type of Formation: Large Boulder
WOW Factor: 4
Location: Pound Ridge (Westchester County)
Tenth Edition, NYS Atlas & Gazetteer: p. 109, D6–7
Earlier Edition, NYS Atlas & Gazetteer: p. 33, C6–7
Parking GPS Coordinates: 41°12.085′N, 73°35.279′W
Henry Morgenthau Preserve Boulder GPS Coordinates: 41°12.018′N, 73°35.313′W (estimated)
Hours: Daily, sunrise to sunset
Accessibility: 0.1–0.2-mile hike
Degree of Difficulty: Moderately easy
Additional Information: Henry Morgenthau Preserve, Route 172, Pound Ridge, NY 10576
Trail Map: henrymorgenthaupreserve.org/trail-map

Description

In the *Eastern New York Chapter Preserve Guide: Lower Hudson Region*, the authors describe the 6-foot-high boulder as "a large glacial erratic (boulder) perched on a ridge."

History

This unnamed glacial boulder is contained in a 36-acre preserve that abuts 45-acre Blue Heron Lake. The initial land that served as a nucleus for the preserve was donated in 1972 by Ruth M. Knight, her daughter, Ellin N. London, and son-in-law, Dr. Robert D. London, in memory of Ruth's father

Henry Morgenthau, a U.S. diplomat, attorney, and real estate investor. Two additional land acquisitions were made in the years that followed.

The preserve was previously affiliated with the Bedford Audubon Society, which acted as the organization's fiscal sponsor.

Directions

From Bedford (junction of Routes 172/Pound Ridge Road & 22), drive east on Route 172/Pound Ridge Road for 3.0 miles and turn right into a small parking area 0.1 mile after passing by Tatomuck Road, on your left. Look for a preserve sign at the park entrance, just after the guardrail ends.

The 6-foot-high boulder is found at the junction of the Blue Trail and White Trail (the latter possibly being unmarked).

Resources

Chris Harmon, Matt Levy, and Gabrielle Antoniadis, *Eastern New York Chapter Preserve Guide: Lower Hudson Region* (Mt. Kisco, NY: Nature Conservancy, 2000), 32–33.

"Henry Morgenthau Preserve." https://www.americantowns.com/place/henry-morgenthau-preserve-katonah-ny.html. The website states that "the largest boulder in the preserve, located at the split of the Blue and White Trails, is over six feet in diameter. It was carried by glacial action thousands of years ago."

"The Henry Morgenthau Preserve." https://www.henrymorgenthaupreserve.org.

Choate Sanctuary Boulder 31

Choate Audubon Sanctuary

Type of Formation: Large Boulder
WOW Factor: Unknown
Location: Mt. Kisco (Westchester County)
Tenth Edition, NYS Atlas & Gazetteer: p. 108, D4–5
Earlier Edition, NYS Atlas & Gazetteer: p. 33, BC5–6
Parking GPS Coordinates: 41°12.537′N, 73°44.797′W
Trailhead GPS Coordinates: 41°12.482′N, 73°44.815′W
Choate Sanctuary Boulders GPS Coordinates: *Main boulder*: Not determined; *Mini-car-sized boulder*: 41°12.436′N, 73°45.020′W
Hours: Daily, dawn to dusk
Accessibility: <0.5-mile walk
Degree of Difficulty: Moderately easy
Additional Information: Choate Audubon Sanctuary, Route 133 & Crow Hill Road, Mount Kisco
Trail Map: sawmillriveraudubon.org/smra-maps/Choate_Sanctuary.pdf
sawmillriveraudubon.org/maps/

Description

Peggy Turco and Katherine S. Anderson, in *Walks and Rambles in Westchester and Fairfield Counties: A Nature Lover's Guide to 36 Parks and Sanctuaries*, mention "a huge erratic—a boulder that stands alone dropped here by a glacier."

Several sources mention unusual rock outcroppings which are formed out of Fordham Gneiss.

History

The 32-acre sanctuary began in 1972 with a 23-acre bequest from the heirs of Joseph H. Choate Jr. to the New Castle Land Conservancy in memory of their father. In 1974, three additional acres were acquired from Geoffrey Platt in memory of his wife, Helen Choate Platt; and then, in 1997, an additional four acres were added under a 99-year lease from the town of New Castle.

The sanctuary has been owned and operated by the Saw Mill River Audubon since 1975.

Directions

From Mt. Kisco (junction of Routes 133/West Main Street & 117/North Bedford Road), drive west on Route 133/West Main Street/Millwood Road for 1.1 miles. Turn right onto Crow Hill Road and head north for 0.1 mile. Then turn right and park along Red Oak Lane, being sure not to block any of the homeowners' driveways.

Walk carefully down Crow Hill Road to reach the sanctuary's entrance, which you passed on the way up to Red Oak Lane. Follow the 0.5-mile-long, white-marked White Oaks Trail that begins at the park's entrance. The boulder is encountered along the trail.

A mini-car-sized boulder is located near the south end of the preserve, about 100 feet up from the highway.

The park's other two trails are the 0.3-mile-long, blue-marked Swamp Loop and the 0.2-mile-long, yellow-marked Hickory Trail

Resources

Peggy Turco and Katherine S. Anderson, *Walks and Rambles in Westchester and Fairfield Counties: A Nature Lover's Guide to 36 Parks and Sanctuaries* (Woodstock, VT: Backcountry Publications, 1993), 67.

"Choate Sanctuary – Saw Mill River Audubon." https://www.sawmillriveraudubon.org/choate/.

Kidd's Rock

32

Kingsland Point Park

Type of Formation: Historic Rock
WOW Factor: Unknown; likely, 1–2
Location: Sleepy Hollow (Westchester County)
Tenth Edition, NYS Atlas & Gazetteer: p. 108, E3
Earlier Edition, NYS Atlas & Gazetteer: p. 33, CD4–5
Parking GPS Coordinates: 41°05.331′N, 73°52.241′W
Kidd's Rock GPS Coordinates: 41°05.050′N, 73°52.417′W (estimated)
Fee: Modest entry fee charged per car
Hour: Daily, 8:00 a.m. to dusk
Accessibility: 0.4-mile walk
Degree of Difficulty: Moderately easy
Additional Information: Kingsland Point Park, 299 Palmer Avenue, Sleepy Hollow, NY 10591

Description

Kidd's Rock is most likely a rock outcropping on the bank of the Hudson River at a spot where it was once possible for a pirate's ship to make landing.

History

Kidd's Rock is supposedly the spot where Captain William Kidd landed while conspiring with Frederick Philipse I to smuggle illicit goods inland. Philipse's role was to light a fire on the rock to guide Kidd's ship in.

Kingsland Point Park is an 18-acre riverfront park at the mouth of the Pocantico River named after Ambrose Kingsland who erected a summerhouse (gazebo) on the rock in the nineteenth century. Kingsland Point is formed at the confluence of the Pocantico River and Hudson

River. Because the point proved hazardous to navigation, a lighthouse was erected near it in 1883.

The park is owned by Westchester County and is operated and maintained by the village of Sleepy Hollow. It opened in 1926.

An old postcard shows a large shoreline boulder (Kidd's Rock) with a gazebo resting on top of it. Today, the gazebo is gone and the boulder is fairly obscured by a seawall that has been built around it.

Captain Kidd seems to have been everywhere. In this book, we recount a number of caves and rocks in downstate New York that are identified with the pirate. One of the citations, however, has eluded us so far. In *A Guide to New Rochelle and Lower Westchester*, Robert Bolton Jr. writes, "On the opposite shore of Long Island is a small jutting promontory, which runs into the entrance of Hempstead Bay, called to this day 'Kidd Rock.'" We can't seem to find any modern references to this promontory. Perhaps it has become part of the Bar Beach Town Park on the west side of the harbor.

There is also a Kidd's Point on the west side of the Hudson River across from the Hudson Highlands. It has gone by other names as well, including Caldwell's Point, Donder Berg Point, and Jones Point (41°17.176'N, 73°57.349'W). Nothing, however, is as colorful as invoking the name of Captain Kidd.

Directions

From Sleepy Hollow (junction of Routes 9/North Broadway & 448/Bedford Road), drive north on Route 9/North Broadway for >0.5 mile Turn left onto Palmer Avenue and proceed west for 0.4 mile, crossing over the Amtrak rail line just before the Hudson River. Turn left and drive south for 0.3 mile to reach the parking area.

From the parking area walk southwest to the end of the park. From what we've read, Kidd's Rock is located not far from the south end of Kingsland Point Park, near the Sleepy Hollow Lighthouse (41°05.043'N, 73°52.455'W). There is little of Kidd's Rock to see other than to enjoy an interesting walk with ambiance and a sense of history surrounding you.

Resources

Henry Steiner, *The Place Names of Historic Sleepy Hollow and Tarrytown* (Bowie, MD: Heritage Books, 1998), 70–71.

Richard M. Lederer Jr., *The Place-Names of Westchester County, New York* (Harrison, NY: Harbor Hill Books, 1978), 77. According to Lederer, the story about Kidd landing at the point has never been confirmed.

Robert Bolton Jr., *A Guide to New Rochelle and Lower Westchester* (Harrison, NY: Harbor Hill Books, 1976 facsimile of a 1842 book), 27.

Wallace Bruce, *The Hudson: Three Centuries of History, Romance and Invention*, centennial ed. (New York: Walking News, 1982), 80–81. Bruce recounts the story of Captain Kidd.

"Kidd's Rock – Sleepy Hollow Country." https://sleepyhollowcountry.com/2022/04/08/captain-kidd/

Rockefeller State Park Preserve Rocks

33

Rockefeller State Park Preserve

Type of Formation: Large Rock; Historic Rock
WOW Factor: 5
Location: Sleepy Hollow (Westchester County)
Tenth Edition, NYS Atlas & Gazetteer: p. 108, E3–4
Earlier Edition, NYS Atlas & Gazetteer: p. 33, CD4–5
Parking GPS Coordinates: 41°06.693′N, 73°50.222′W
Destination GPS Coordinates: *Glacial Erratic*: 41°06.511′N, 73°50.562′W (estimated); *Spook Rock*: 41°06.476′N, 73°51.211′W (estimated); *Raven Rock*: 41°05.800′N, 73°48.966′W (per Google Earth)
Fee: Entrance fee charged
Hours: Daily, dawn to dusk
Accessibility: *Glacial Erratic*: 0.5-mile hike; *Spook Rock*: 1.5-mile hike; *Raven Rock*: Probably >2.0-mile hike. Bring along a trail map in order to navigate the complexity of the trail system.
Degree of Difficulty: *Glacial Erratic*: Moderate; *Spook Rock*: Moderate; *Raven Rock*: Moderately difficult
Additional Information: Rockefeller State Park Preserve, 125 Phelps Way, Pleasantville, NY 10570
Trail Maps: parks.ny.gov/documents/parks/RockefellerTrailMap.pdf
friendsrock.org/rspp-map

Description

Glacial Erratic is just as the name suggests—a large rock that is not indigenous to the immediate area. Atlas Obscura writes, "Standing at the base of the 20-foot-high glacial erratic in Rockefeller State Park Preserve, one can't help to notice the sweeping grey, blue, and brown striations that wind their way around this boulder's 65-foot circumference; often interrupted by jagged edges and deep gouges." The boulder is estimated to weigh 8.5 tons. Benches are lined up in front of the rock to create an outdoor classroom. As part of "I love my park day," new benches were installed in 2019 by a group of kids.

In *Day Walker: 32 Hikes in the New York Metropolitan Area*, the authors write, "Reputedly, this stone is the largest such boulder deposited in this part of the country by the receding glacier." Given the fact that there are some really big glacial boulders in this part of the country, we are assuming that the writers meant to say "county" instead of "country."

Spook Rock is described by one source as a "large rock." We suspect that it is associated with a bluff that may also be called Spook Rock. In *The Place Names of Historic Sleepy Hollow and Tarrytown*, for instance, Henry Steiner describes Spook Rock as an "unmarked . . . large flat rock."

Raven Rock is a massive rock or group of rocks that were detached from the cliffs on the east side of Pocantico Hills east of Ferguson Lake. According to a quote from the *Westchester Historian: Quarterly of the Westchester County Historical Society* in Henry Steiner's *Place Names of Historic Sleepy Hollow and Tarrytown*, "It is an unusual arrangement of rocks with perpendicular walls, deep crevices and an old cave that has almost completely disappeared."

Raven Rock has also been known as Crow's Rock.

History

The Rockefeller State Park Preserve is a >1,771-acre park established as a state park in 1983. It was donated to New York State by the William Rockefeller family in 1963. The 45 miles of crushed stone carriage roads were laid during the first half of the twentieth century by John D. Rockefeller Sr. and John D. Rockefeller Jr. The landscape was designed by Frederick Law Olmsted.

The park is presently managed by the New York State Office of Parks, Recreation, and Historic Preservation.

For those who are interested in cinematic information, the park's entrance on Route 117 was used briefly in the 2002 movie *Super Troopers*.

SPOOK ROCK

According to Native American folklore, a young man spied twelve beautiful, young girls dancing by Spook Rock. He kidnapped the loveliest of the twelve and took her home to be his bride. They had a baby together, but all was not to be well. Within three years, the baby had died, and then the mother. It is said that the wandering ghost of this young mother still returns to the rock, looking for her husband and baby.

Strange lights have been reported in the vicinity, presumably caused by the spirit of the young mother. A nice Halloween story for Spook Rock, if you believe in this kind of thing.

The name Spook Rock is not as uncommon as you might at first think. Just in eastern New York State alone there are spook rocks in Montebello, Suffern, and Greenport Center.

RAVEN'S ROCK

A legend accompanies this rock. An unfortunate woman became lost in a snowstorm and sought shelter between the rock and the hillside. She ended up dying from exposure. It is rumored that her ghost now rises whenever someone approaches in the winter to warn them of impending danger. In the first volume of *A History of the County of Westchester, from Its First Settlement to the Present Time*, Robert Bolton Jr. writes that the rock "is now haunted by . . . the lady in white, whose shrill shrieks are said to be often heard during the long and weary winter nights, as if presaging a storm. Tradition asserts that she perished here in deep snow."

Richard M. Lederer Jr., in *The Place-Names of Westchester County, New York*, describes it as a "semi-circular rock formation."

Directions

From north of Philipse Manor (junction of Routes 117/Phelps Way & 9/ Broadway), drive northeast on Route 117/Phelps Way for 1.3 miles and turn right into the park's entrance, parking by the visitor center.

GLACIAL ERRATIC

From the parking lot, follow the Old Sleepy Hollow Road Trail (an old colonial road) southwest. By 0.4 mile, you will come to Nature's Way Path on your right. Follow it north for a short distance, and then turn left onto a spur path that takes you quickly to the glacial boulder.

SPOOK ROCK AND RAVEN ROCK

The directions to these rocks are complex enough that you will need to pick up a trail map when you go to the preserve. Spook Rock is located slightly south of Route 117/Phelps Way. Raven Rock is near the east side of the park, roughly midway along the north/south running border.

By good fortune, all three sites are shown on the Rockefeller State Park Preserve trail map.

Resources

New York-New Jersey Trail Conference, *Day Walker: 32 Hikes in the New York Metropolitan Area*. 2nd ed. (Mahwah, NJ: New York-New Jersey Trail Conference, 2002), 139, 141. A trail map is on page 136.

Richard M. Lederer Jr., *The Place-Names of Westchester County, New York* (Harrison, NY: Harbor Hill Books, 1978), 118.

"Eagle Hill Area – Rockefeller State Park Preserve." https://scenesfromthetrail.com/2017/03/18/eagle-hill-area-rockefeller-state-park-preserve/. This site contains photographs of the glacial erratic.

Henry Steiner, *The Place Names of Historic Sleepy Hollow and Tarrytown* (Bowie, MD: Heritage Books, 1998), 132. Steiner recounts the legend behind Spook Rock.

William Owens, *Pocantico Hills 1609–1959* (Sleepy Hollow, NY: Sleepy Hollow Restorations, 1960), 2. A photograph of Raven Rock is shown along with a caption narrating the rock's legend.

Robert Bolton Jr., *A History of the County of Westchester, from Its First Settlement to the Present Time*, vol. 1 (NY: Alexander S. Gould, 1848), 439.

"Glacial Erratic – Pleasantville, New York – Atlas Obscura." https://www.atlasobscura.com/places/glacial-erratic.

Helicker's Cave and Big Boulders 34

Betsy Sluder Preserve

Type of Formation: Rock-Shelter; Boulder
WOW Factor: 6
Location: Armonk (Westchester County)
Tenth Edition, NYS Atlas & Gazetteer: p. 108, E5
Earlier Edition, NYS Atlas & Gazetteer: p. 33, CD6
Parking GPS Coordinates: 41°07.239′N, 73°43.025′W
Destination GPS Coordinates: *Helicker's Cave*: 41°07.020′N, 73°43.493′W (estimated); *Big Boulders*: 41°07.407′N, 73°43.484′W
Accessibility: *Helicker's Cave*: 0.7-mile hike; *Big Boulders*: Not determined; *Perimeter of Preserve*: 1.5-mile hike
Degree of Difficulty: Moderately easy
Trail Map: leathermansloop.org/2009/02/how-do-you-train-betsy-sluder-preserve.

Description

Helicker's Cave is a small-to-medium-sized rock-shelter located on a hillside.

Of Big Boulders, the alltrails.com/trail/us/new-york/betsy-sluder-nature-trail website states that "in the middle of the park there are big boulders that are great fun to climb on." The rocks must be of fairly good size if you can climb around on them.

The exact location of the boulders is not depicted on the park map, but these rocks are definitely worth making an effort to locate. We have come up with a possible GPS reading using Google Earth. We suspect that there will be no difficulty locating these rocks once you are on the trail and fully involved in the hunt.

18. Big rock at Betsy Sluder Preserve. Photograph by Alex Smoller.

History

Helicker's Cave is historically significant, for not only was it occupied by early Native Americans, but reputedly later by Anne Hutchinson and her children, who lived at the site temporarily sometime following the Revolutionary War; then by Old Bet Helicker, a hermit who resided at the rock-shelter from 1783 to 1802 (although one site suggests that Bet Helicker was a "li'l ol' lady" who lived in a shack near the cave); and lastly by the mysterious Leatherman, who used it as one of his stopovers while making his regular 34-day, 365-mile, near monthly circuit. Some of the information about Bet Helicker may not be true, however. In volume 4, number 2 of *The Westchester Historian: Quarterly Bulletin of the Westchester County Historical Society*, the author writes, "Bet Helicker herself did not live in the cave, but in a hut in the woods nearby." The author however, recognize that white men did live in the cave several centuries ago: "In its ashes were found bones of domestic animals bearing saw marks, pieces of clay pipes, and broken pottery of the Revolutionary period."

The Betsy Sluder Preserve is a 70-acre parcel of land in a forested area completely surrounded by homes and businesses. In 1999, the preserve was renamed for Betsy C. Sluder, who was an ardent regional conservationist, involved in many boards and projects.

Google Earth still identifies the preserve as Whippoorwill Ridge Park, which undoubtedly was the park's earlier name. Some older websites refer to the park as Whippoorwill Park, minus the word "Ridge."

According to Richard M. Lederer Jr. in *The Place-Names of Westchester County, New York*, there is also a Little Bet Helicker's Cave, south of Helicker's Cave. More to look for, it would seem.

Directions

From I-684 east of Armonk, take Exit 3 South and head southwest on Route 22/Armonk-Bedford Road for 0.6 mile. At the traffic light, turn right onto Main Street, which takes you north towards the center of Armonk. Within 0.1 mile, turn left onto Old Route 22, and proceed southwest for 0.2 mile.

Park to your right immediately after Birdsall Farm Drive in a small area for the preserve. Check the kiosk for pertinent information.

A 1.5-mile-long trail leads around the perimeter of the preserve. From the parking area, take the red-marked path clockwise west (left), then south, for 0.7 mile. The Leatherman Cave is located just off the trail at the southernmost part of the preserve. A path leads to it.

The grouping of large, moss-covered boulders are located near the middle of the park on a fairly steep hillside.

Resources

Patricia Edwards Clyne, *Caves for Kids in Historic New York* (Monroe, NY: Library Research Associates, 1980), 9–16.

"The Legend of the Leatherman." https://leathermansloop.org/2009/02/the-legend-of-the-leatherman.

Richard M. Lederer Jr., *The Place-Names of Westchester County, New York* (Harrison, NY: Harbor Hill Books, 1978), 17.

Patricia Edwards Clyne, *Hudson Valley Trails and Tales* (Woodstock, NY: Overlook Press, 1990), 82, 162.

"How Do You Train? Betsy Sluder Preserve." https://leathermansloop.org/2009/02/how-do-you-train-betsy-sluder-preserve.

"Betsy Sluder Nature Trail Map Guide." https://alltrails.com/trail/us/new-york/betsy-sluder-nature-trail.

"Whippoorwill Park Woodland Walks." https://woodlandwalks.org/whippoorwill-park/

"Bet Helicker's Cave," *Westchester Historian: Quarterly Bulletin of the Westchester County Historical Society* 4, no. 2 (April 1928): 40. On page 40 also is a photograph of Bet Helicker's Cave.

"The Old Leatherman Caves Guide." https://theairlandandsea.com/2019/04/the-old-leatherman-caves-guide.html.

Horace Sarles Rock and Shelter Caves 35

Wampus Pond County Park

Type of Formation: Large Rock or Rock Outcropping; Rock-Shelter
WOW Factor: Unknown
Location: Armonk (Westchester County)
Tenth Edition, NYS Atlas & Gazetteer: p. 108, E5
Earlier Edition, NYS Atlas & Gazetteer: p. 33, CD6
Parking GPS Coordinates: 41°08.836′N, 73°43.678′W
Destination GPS Coordinates: *Horace Sarles Rock*: 41°08.864′N, 73°43.885′W (guesstimate); *Shelter Cave(s)*: Unknown, but probably in the same general area as Horace Sarles Rock
Accessibility: <0.6-mile bushwhack/hike clockwise around shore of pond (if not posted); >0.1-mile paddle across pond
Additional Information: Wampus Pond County Park, 1 Wampus Lake Drive, Armonk, NY 10504

Description

Horace Sarles Rock is a fairly nondescript rock or rock outcropping that juts out into Wampus Pond.

Rock-Shelter(s): We do not have any specific descriptions about the shelter caves but assume that they are fairly small and likely formed by overhanging ledges on the west side of Wampus Pond.

History

According to Richard Lederer, in *The Place-Names of Westchester County, New York*, Horace Sarles Rock acquired its name when Horace Sarles (most likely a local resident), "was killed by a falling tree while cutting timber on the rock that juts into Wampus Lake."

"Wampus" is Native American for "opossum."

Wampus Pond was earlier known as Wampus Lake Reservoir, a natural body of water, when it was part of the New York City water supply. This ended in 1963 when the pond, including 93 acres, was purchased by the county from New York City.

ROCK-SHELTERS

Parker Harrington, doing a cultural resource inventory of the pond area, found several early twentieth-century rock-shelters in or near the park. We have not been able to find any further information regarding the subject.

In *New York Walk Book*, however, Raymond H. Torrey, Frank Place. Jr., and Robert L. Dickinson do write that "The scenery [along the west shore] is picturesque with high ledges frowning under hemlock forest." This sounds like the ideal spot to find rock-shelters.

Directions

From Armonk (junction of Routes 128/Armonk Road & 22/Armon Bedford Road), drive north, then northwest, on Route 128/Armonk Road for 2.5 miles and turn left into a small parking area by the pond.

The perimeter of the pond is roughly 1.0 mile long, but not all of it needs to be explored.

HORACE SARLES ROCK

We suspect that the rock is on the west side of the pond, probably somewhere along 0.4 mile of shoreline.

ROCK SHELTER

Topographic maps show that the land along the west side of the pond rises up steeply, gaining nearly 250 feet of elevation in 0.1 mile. You couldn't hope for better conditions than this to look for rock-shelters in the park.

Resources

Richard M. Lederer Jr., *The Place-Names of Westchester County, New York* (Harrison, NY: Harbor Hill Books, 1978), 69.

"Wampus Pond." https://parks.westchestergov.com/wampus-pond.

Raymond H. Torrey, Frank Place Jr., and Robert L. Dickinson, *New York Walk Book*, 3rd ed. (New York: American Geographical Society, 1951), 106–107.

Irvington Woods Rocks 36

Irvington Woods

Type of Formation: Split Rock
WOW Factor: 5–6
Location: Irvington (Westchester County)
Tenth Edition, NYS Atlas & Gazetteer: p. 111, A6
Earlier Edition, NYS Atlas & Gazetteer: p. 33, D4–5
Parking GPS Coordinates: 41°02.519′N, 73°50.781′W
Destination GPS Coordinates: *Split Rock*: 41°02.309′N, 73°50.883′W; *Kiosk Boulder*: 41°02.519′N, 73°50.793′W (estimated); *Irving Rock*: 41°02.214′N, 73°51.109′W; *Sunset Rock*: Not determined; *Jenkins Rock*: Not determined
Hours: Daily, dawn to dusk
Accessibility: *Kiosk Boulder*: 0.0-mile walk; *Split Rock*: >0.5-mile hike; *Irving Rock*: Distance not determined; *Sunset Rock*: >0.8-mile hike
Degree of Difficulty: Moderate
Additional Information: O'Hara Nature Center, 170 Mountain Road, Irvington, NY 10533 (914) 591-7736
Trail Map: theirvingtonwoods.org/peter-k-oley-trail-network/peter-k-oley-trail-maps

---

Description

Kiosk Boulder is a big 6-foot-high boulder that stands next to the kiosk at the start of the trail.

Split Rock is a large glacial erratic that, many years ago, broke into two pieces that now lie apart on a slightly rounded mound of bedrock. Both pieces are over 6 feet in height. The larger of the two of is 10 feet in length. The smaller, more compact half, lies tilted on the bedrock, propped up by a smaller rock.

Jenkins Rock is located on the northwestern shore of the Irvington Reservoir and dedicated in 1984 to Rev. Dr. Frederick Jenkins who, while pastor of the Irvington Presbyterian church, conducted Easter Sunday sunrise services overlooking the reservoir.

Sunset Rock is a rocky section of bedrock with views of the Saw Mill River valley that appear during the winter when the trees are bereft of leaves. It is not a boulder.

History

The Irvington Woods contain 400 acres of land. Historical rumor has it that Washington Irving wrote some of his works here.

The O'Hara Nature Center is a recent addition to the woods, having opened in 2012.

The Peter K. Oley trail system is named for Peter K. Oley, who was instrumental in establishing the trail system in the area he called his "back forty."

The Irvington Reservoir dates to 1900 and was initially used as a source of uncontaminated, household water for residents of Irvington.

Interestingly, MapQuest shows the name of the woods as Fieldpoint Park.

Directions

Driving southwest on the Saw Mill River Parkway, go under the New York State Thruway (I-87) and immediately turn sharply right onto Mountain Road. Watch out for incoming traffic from the Thruway as you make your turn. Head southwest on Mountain Road for 0.8 mile and turn left into the parking area for the O'Hara Nature Center.

KIOSK BOULDER

Look for the boulder near the kiosk close to the Nature Center's parking area.

SPLIT ROCK

From the Nature Center, follow the North-South (NS) trail west, then south. When you come to the junction with the Split Rock Trail (SR), turn right and go west. Look for Split Rock on your right.

JENKINS ROCK

This historic rock is located at the northwest end of the Irvington Reservoir along the Hermit's Grave Trail.

SUNSET ROCK

Continue south on the North-South Trail (NS) until you come to the Sunset Rock Trail (SN). Turn left and head east. Look for Sunset Rock on your left. Just remember—this is an overlook, not a boulder.

ALTERNATE PARKING

At the southeast end of the Irvington Reservoir, off Cyrus Field Road, is a small parking area [41°02.068′N, 73°50.916′W]. The road was named for Cyrus Field, an American businessman who, along with others in the Atlantic Telegraph Company, laid the first telegraph cable across the Atlantic Ocean in 1858.

Resources

"The Irving Woods." https://www.theirvingtonwoods.org.

"Irving Woods." https://scenesfromthetrail.com/2017/06/17/irvington-woods. This website contains a photograph of Split Rock.

"The Peter K. Oley Trail Network." https://www.theirvingtonwoods.org/peter-k-oley-trail-network/.

Aquehung Boulder 37

Type of Formation: Large Boulder
WOW Factor: 5
Location: Bronxville (Westchester County)
Tenth Edition, NYS Atlas & Gazetteer: p. 111, B6–7
Earlier Edition, NYS Atlas & Gazetteer: p. 25, A5
Aquehung Boulder GPS Coordinates: 40°55.915'N, 73°50.991'W
Accessibility: Roadside. Respect private property and just look from the road.
Degree of Difficulty: Easy

Description

An old photograph shows a big rock tilted at a 45-degree angle. The rock is virtually as tall as the two-story townhouses that abut it on both sides, and is probably well over 15 feet in height.

It is gratifying to see that developers allowed the rock to stand in place as they built around it.

History

According to Frank L. Walton, in *Pillars of Yonkers*, the "history [of the rock] is unknown but probably dates back to the Glacial Age. At least one of the oldest objects in Yonkers."

The word *aguehung* is Mohican for "River of High Bluffs," a nod to the Bronx River and one section where the banks rise up to as high as 75 feet.

Directions

The mammoth rock is located on Midland Avenue just 0.1 mile northwest of Midland Avenue's junction with Wrexham Road. The boulder stands between two townhouse complexes, 50 feet from the road. A sign near the road reads: "Judge Arthur J. Doran Townhouses."

There is no easy way to give directions to this site. If you happen to be on I-87, just north of where it goes under the Cross County Parkway, weave over to your right onto Central Park Avenue and take the exit for Midland Avenue. When you come up to Midland Avenue, turn left and head southeast for <0.4 mile. The rock will be on your left.

Otherwise, we would suggest obtaining directions for Sarah Lawrence College or Sunnybrook Park and, from there, plot your way over to Midland Avenue.

Resources

Frank L. Walton, *Pillars of Yonkers* (New York: Stratford House, 1951), 237. A photograph of the rock can be seen on an insert between pages 274 and 275.

"The Broncs and Harlem," New York City Forum. https://www.tripadvisor.com/ShowTopic-g60763-i5-k1974928-The_Broncs_and_Harlem-New_York_City_New_York.html.

Anita Inman Comstock, *Wondrous Westchester: Its History, Landmarks, and Special Events* (Mount Vernon, NY: Effective Learning, 1984), 43.

Farcus Hott Cave 38

Type of Formation: Rock-Shelter
WOW Factor: 3
Location: Elmsford (Westchester County)
Tenth Edition, NYS Atlas & Gazetteer: p. 111, A7
Earlier Edition, NYS Atlas & Gazetteer: p. 33, D5
Parking GPS Coordinates: 41°03.454′N, 73°49.477′W
Farcus Hott Cave GPS Coordinates: 41°03.010′N, 73°49.717′W (estimated)
Accessibility: 0.7-mile hike
Degree of Difficulty: Moderate

Description

According to Rob Yasinsac on his website, hudsonvalleyruins.org/rob/?p=541, this historic cave "appears to be a pile of boulders massed together from some great natural calamity, all tossed about in disarray around a small, mouth-shaped opening under a ledge." In other words, what sounds like a talus cave.

History

Farcus Hott Cave, also known as Kathy's Cave, was recently rediscovered (presumably, in name only) by Lucas Buresch. We say this because the cave is very close to several housing developments.

Legend has it that the cave was used as a hideout by colonial farmers to avoid capture from British soldiers sweeping across the landscape.

Directions

From the Cross Chester Expressway, take Exit 1 for Route 119/Tarrytown/Saw Mill Parkway. At the end of the ramp, turn left onto Route 119/West Main Street and head southeast for <0.3 mile. Park to your left in a large, paved area (currently next to the Eldorado Diner) crossed over by power lines.

Walk carefully across West Main Street and then follow an old dirt road, if not posted, for 0.7 mile along a power line corridor as you ascend Beaver Hill.

When you begin descending Beaver Hill, still following the power line corridor, look for the rock-shelter next to the left side of the corridor. (We know this because in a photograph taken from inside the rock-shelter, the power lines are visible to your right.)

You are not that far from the Saw Mill River Parkway, which lies directly downhill from the shelter cave.

The entire area is circumscribed by the Saw Mill River Parkway to your east and south, I-87 to your west, and I-287 to your north.

It's worth keeping in mind that there are housing apartments (Avalon Green Apartments and Ridgeview Apartments) nearby, which means it might be possible to reach the power line corridor with only minimal effort if you start from near the top of the hill. Once again, if the land beyond the apartments is posted, then go no further.

Resources

"Farcus Hott (Katy's Cave), Greenburgh, NY." www.hudsonvalleyruins.org/rob/?p=541. This website tells the story about how a group of modern day explorers sleuthed out the location of this cave.

Lucille and Ted Hutchinson, *Storm's Bridge: A History of Elmsford, N.Y., 1700–1976*, (Elmsford, NY: Bicentennial Committee, 1980), 33.

"The Search for Farcus Hott – Archive Sleuth." https://archivesleuth.wordpress.com/2012/02/02/the-search-for-farcus-hott/.

Sigghes Rock and Amackassin Rock (Historic) 39

Type of Formation: Large Rock
WOW Factor: Unknown
Location: Yonkers (Westchester County)
Tenth Edition, NYS Atlas & Gazetteer: p. 111, B6
Earlier Edition, NYS Atlas & Gazetteer: p. 25, A4–5
Andrus Foundation GPS Coordinates: 40°59.032′N, 73°52.790′W
Street Addresses Associated with Amackassin Rock—
GPS Coordinates: *Amackassin Terrace*: 40°57.239′N, 73°53.108′W; *Warburton Avenue*: 40°57.405′N, 73°53.691′W

Description

Many years ago, two rocks that were located near opposite ends of Amackassin Creek, aka Meccackassin Creek, achieved a modicum of notoriety. One was called Sigghes Rock; the other was called Amackassin Rock.

Sigghes Rock marked the Yonkers-Hastings line, and earlier was a significant Indian boundary stone. Judging from a photograph in George L. McNew's article, "The Paradise World of the Red Man in Westchester County," the rock may be at least 8 feet high and many times as long. In his article on "Sigghes Rock," Frank L. Walton writes, "5,000 years ago an old animal trail passed near the rock. The Indians discovered it some 3,000 years ago . . . the ancient trail has become the Albany Post Road." As an aside, one has to wonder how Walton knew so precisely that an old animal path had passed by the rock 5,000 years ago!

According to *Postscript*, an online magazine, "Sigghes [rock] stands on the Andrus Foundation property and marks the boundary between Greenburgh and Yonkers."

The Amackassin Rock marked the extreme northwestern corner of Yonkers. It was a large rock, but old photographs don't give a clue as to its actual size. According to Walton, the glaciers carried "a great Copper Colored Rock [Amackassin Rock] which it deposited in the waters of the Hudson River, some 25 feet from the shore. This rock was worshipped by the Indians because it sparkled in the sun and glowed in the moonlight."

The online magazine *Postscript* states that "Amackassin [is a] large rock on the shore of the Hudson River. It marked the northern boundary of Van der Donck's purchase."

Walton gives two different accounts of what may have happened to Amackassin Rock. The first account contends that the rock was destroyed in 1848 when the Hudson River Railroad was built along the edge of the river. The second account advances the notion that the rock was never actually destroyed, but merely buried under rubble when the railroad line was built. If so, then this historic rock still exists, albeit now underground.

The third possibility, which photographer Travis Paige advances, is that the boulder still exists and lies above ground. He has taken photographs of it, assuming, of course, that this large boulder is actually the historic Amackassin Rock.

According to an article called "The Amackassin Stone," "the name of the rock is likely derived from two Delaware Indian words, '*mackaak*' (great) and '*acksin*' (stone)."

History

Amackassin Creek is (or was) a small, 0.5-mile-long brook that flowed from the North Broadway hillside to the Hudson River.

Directions

As far as we can tell, Amackassin Creek no longer exists, or else is so insignificant that it doesn't show up on local maps. Perhaps the creek was incorporated into an underground sewer/water drainage line like so many of the creeks have been in Albany, a city where Barbara Delaney (my wife and hiking companion) and I live.

The lack of recognizable landmarks makes finding the site of Amackassin Rock difficult for sure. However, one substantial clue is offered by Travis Paige on his website Travis Paige Photography (travispaigephotography.blogspot.com/2015/). Amackassin Rock "can be viewed by parking at the top off Wolf Run and taking a quick walk on the trail behind the garbage dumpsters staying to your right." Paige goes on to mention that "there are also some other large boulders nearby."

Sigghes Rock, on the other hand, seems to be within the realm of possibility to find. According to the January 1967 issue of the *Yonkers Historical Bulletin*, Sigghes Rock lies on the east side of Broadway on the Yonkers-Hasting boundary line. Richard M. Lederer Jr., in *The Place-Names of Westchester County, New York* states that the rock is on the Andrus Foundation property, marking the Greenburg-Yonkers boundary.

We tried to figure out where the Yonkers/Hasting-on-Hudson border crosses the Andrus Foundation property, and it seems to be somewhere around 0.3 mile south of the main building. The rock's exact location, however, continues to elude us. If you go to look for it, be sure to check first with the staff at the Andrus Foundation for permission before setting off.

Resources

George L. McNew, "The Paradise World of the Red Man in Westchester County," *Yonkers Historical Bulletin* 15, no. 1 (January 1965). On page 11 is an old photograph of Sigghes Rock.

"Legendary Amackassin Rock Shown in Photograph," *Yonkers Historical Bulletin* 9, no. 1 (April 1962): 16.

Frank L. Walton, "Sigghes Rock," *Westchester Historian: The Quarterly of the Westchester County Historical Society* 42, no. 3 (Summer 1966): 47.

"The Amackassin Stone," *The Westchester Historian: Quarterly Bulletin of the Westchester County Historical Society* 5, no. 1 (January 1924): 6–7.

Edward Martin (photographer), *Yonkers Historical Bulletin* 9, no. 1 (April 1962). The caption reads, "Legendary Amackassin Rock shown in photograph." The text, not ascribed to an author, states that the photograph was taken by Edward Martin in 1940.

Yonkers Historical Bulletin 15, no. 1 (January 1968). On page 11 is a photograph of Sigghes Rock.

Robert Bolton Jr., *A History of Westchester from Its First Settlement to the Present Time*, vol. 2 (New York: Alexander S. Gould, 1848). On page 404 is a line drawing of Amackassin Rock.

"Travis Paige Photography: 2015 – Blogger." https://travispaigephotography.blogspot.com/2015/.

"Hastings Historical Society – Sigghes or Sigges Rock." https://www.facebook.com/HastingsHistoricalSociety/photos/sigghes-or-sigges-rock-in-the-andrus-meadow-on-the-line-between-yonkers-and-hast/10158102510360098/. This site contains a photograph of the large rock.

Benziger Boulder 40

Type of Formation: Large Rock
WOW Factor: 4–5
Location: Bronxville (Westchester County)
Tenth Edition, NYS Atlas & Gazetteer: p. 111, B6–7
Earlier Edition, NYS Atlas & Gazetteer: p. 25, A5
Parking GPS Coordinates: 40°56.704′N, 73°49.963′W
Benziger Boulder GPS Coordinates: 40°56.739′N, 73°50.001′W
(per bouldering website)
Accessibility: 0.1-mile hike
Degree of Difficulty: Moderately easy

Description

Benziger Boulder is a 15-foot-high rock that was presumably named for a rock climber or boulderer named Benziger.

Directions

From Bronxville (junction of Sagamore Road and Avon Road), drive southwest on Sagamore Road for >0.1 mile and park to your right, just down the road from the Sagamore Road Playground.

You may need to consult a map to determine how best to get to Bronxville and, from there, to Sagamore Road.

Park along the side of Sagamore Road where parking is free for two hours.

Then follow a path that leads downhill through the park to the northwest quadrant. The rock is located near Kensington Road. It is hidden a bit from view, so you may have to look for it.

Resources

"Rock Climbing Routes in Westchester County, New York." https://rockclimbing.com/routes/North_America.United_States/New_York/Westchester_County/Benziger_Boulder.

Indian Rock Shelter 41

Type of Formation: Rock-Shelter
WOW Factor: 7–8
Location: Larchmont (Westchester County)
Tenth Edition, NYS Atlas & Gazetteer: p. 111, B8
Earlier Edition, NYS Atlas & Gazetteer: p. 25, A5–6
Indian Rock Shelter GPS Coordinates: 40°55.710′N, 73°45.767′W (estimated)
Accessibility: Unknown

Description

Judith Doolin Spikes, in *Larchmont, NY: People and Places*, writes that the Indian Rock Shelter has "a deep recess under it that could easily be made into a large and well-sheltered living room." A photograph that we have seen shows a huge slab of rock leaning against the side of a hill or slope, forming a tall, enterable shelter. In "The Indian Rock Shelter of Larchmont," Morgan H. Secord writes that the rock "was found to have a very deep undercut causing a very high and wide overhang and thereby a deep recess under it that could easily be made into a large and well-sheltered living room by the addition of a sloping enclosure formed of poles and tree trunks, with a covering of the bark of trees of the outside." (See chapter on "Bowdoin Park Rockledge Shelter" for a picture of what this would look like).

History

The rock-shelter was first identified as being on a tract of land sold to Joseph Secord in 1794 by John Bailey. At the time, some imaginative onlookers were able to see the face of a Native American in the rocks, but by the 1960s the facial profile had apparently disappeared or been obscured.

Directions

The 1962 issue of the *Westchester Historian: Quarterly of the Westchester County Historical Society* states that the rock is "located a few yards south of Palmer Avenue, a short distance west of the last of the series of stores on that street. It adjoins a dry bed of Gravely Brook (now confined in a culvert)."

Much has changed over the last 50 years. Larchmont is now heavily populated and Palmer Avenue particularly so. The rock-shelter may no longer exist or could be squirreled away behind someone's house.

To get to Pine Brook Park as a starting point, from near the center of Larchmont (junction of Chatsworth Avenue & Route 1/Boston Post Road), drive northwest on Chatsworth Avenue for 0.5 mile. Turn left onto Palmer Avenue and head southwest for 0.3 mile. Pine Brook Park is directly to your left at the intersection of Palmer Avenue and Pine Brook Drive. Park along the south side of Palmer Avenue just east of Pine Brook Drive.

From a photograph we have seen, we believe Indian Rock is located at the south end of the park. What's frustrating is that we cannot see it on Google Earth. Hopefully, the rock is at the suggested location.

Resources

Morgan H. Secord, "The Indian Rock Shelter of Larchmont," *Westchester Historian of the Westchester County Historical Society* 38, no. 2 (April–June 1962): 49–50. Officer Dix Bruggese's photograph of the rock shelter is on page 48.

"Pine Brook Park Renovation Underway in Larchmont." https://patch.com/new-york/larchmont/pine-brook-park-renovation-underway-in-larchmont.

Judith Doolin Spikes, *Larchmont, NY: People and Places* (Larchmont, NY: Fountain Square Books, 1991). In the preface, page unnumbered, can be seen a photograph of the rock. The accompanying article is entitled "Indian Rock Shelter or Profile Rock?"

Rocking Stone 42

Type of Formation: Large Rock
WOW Factor: 7
Location: Larchmont (Westchester County)
Tenth Edition, NYS Atlas & Gazetteer: p. 111, B7–8
Earlier Edition, NYS Atlas & Gazetteer: p. 25, A5–6
Rocking Stone GPS Coordinates: 40°56.412′N, 73°46.067′W
Accessibility: Roadside
Degree of Difficulty: Easy

19. Larchmont Rocking Stone. Antique postcard, public domain.

Description

Rocking Stone is an 11-foot-high, 150-ton glacial erratic. Reputedly, the boulder was so perfectly balanced at one time that it behaved like a seesaw and could be rocked back and forth. In 1842, Robert Bolton Jr., in *A Guide to New Rochelle and Lower Westchester*, wrote that "this natural curiosity is a mass of solid rock, weighing, perhaps, more than twenty tons, which can be moved to and fro, at pleasure, by a child." That was then—nearby blasting by workers in the 1920s to put in a sewer line destabilized the rock sufficiently to make it inert.

History

The curious engravings visible on the rock are explained in Judith Doolin Spikes's *Larchmont, NY: People and Places*. "In 1854, the Chatsworth Land Company hired a New Rochelle engineer, William Bryson, to survey their land and draw a map. When Bryson finished the survey, he carved his name, the name of the company, and the date into the Rockingstone."

According to Richard M. Lederer Jr., in *The Place-Names of Westchester County, New York*, "The site of the rockingstone on Rockingstone Avenue once bore the name [Hannah's Peak]. Her [Hannah's] last name is unknown."

In 2012, a filmmaker who wanted to use the boulder in a forthcoming movie walked into the Mamaroneck town administrator's office and asked if he could obtain a temporary permit to restore the boulder to its original condition again. The request was promptly denied. Apparently, it's just not in the cards for the Rocking Stone to rock and roll again—ever. This is probably just as well since the rock, due to its prominent location, would prove dangerously irresistible to locals wishing to do mischief.

Directions

The Rocking Stone is located at the center of the intersection of Rockingstone Avenue, Spruce Road, Poplar Road, and Springdale Road in Larchmont.

To get there from Route 1/Boston Post Road, take Chatsworth Avenue northwest for 1.0 mile. As Chatsworth Avenue veers right, continue straight ahead on Rockingstone Avenue for another 0.2 mile to reach the boulder.

Resources

Judith Doolin Spikes, *Larchmont, NY: People and Places* (Larchmont, NY: Fountain Square Books, 1991), 57.

Robert Bolton Jr., *A Guide to New Rochelle and Lower Westchester* (Harrison, NY: Harbor Hill Books, 1976 facsimile of the 1842 book), 40–41.

"He Wants to Rock the Rocking Stone." https://theloopny.com/he-wants-to-rock-the-rocking-stone-a-larchmont-landmark.

Herbert B. Nichols, *Historic New Rochelle* (New Rochelle, NY: Board of Education, 1938), 112.

Richard M. Lederer Jr., *The Place-Names of Westchester County, New York* (Harrison, NY: Harbor Hill Books, 1978), 122.

Manor Park Rocks 43

Larchmont Manor Park

Type of Formation: Large Rocks
WOW Factor: 3
Location: Larchmont (Westchester County)
Tenth Edition, NYS Atlas & Gazetteer: p. 111, B8
Earlier Edition, NYS Atlas & Gazetteer: p. 25, A5–6
General GPS Coordinates for Manor Park: 40°55.048′N, 73°44.904′W
Destination GPS Coordinates: *Dinosaur Egg*: 40°55.052′N, 73°44.680′W (estimated); *Other rock features*: Not determined
Accessibility: 1.0-mile walk along path at top of rocky shoreline
Degree of Difficulty: Moderately easy

Description

A number of medium-to-large-sized rocks can be seen along the pathway overlooking the Long Island Sound.

Sliding Rock is a granite erratic whose side was rasped by the action of glaciers. It is located near the pumping station.

Whale-Back is found near the flagpole. It got its rounded hump as it was rolled by glaciers.

Death's Head lies opposite the flagpole, near the shore, balanced on native rock.

Dinosaur Egg is located roughly midway between the South and North Gazebos.

History

Manor Park encompasses 13 acres of land, with nearly a stretch of 1.0 mile along the Long Island Sound and Larchmont Harbor. It is privately owned but open to the public.

Land ownership goes back to 1661, when John Richbell purchased three necks of land from Native Americans. In 1701, Samuel Palmer purchased the middle neck, which consisted of land encompassing Larchmont and Larchmont Manor Park. Most of this land was subsequently purchased by Peter Jay Munro (nephew of John Jay). In the 1870s, the initial 6.0 acres for Manor Park was established by land developer Thomas J. S. Flint.

The Larchmont Manor Park Society was formed in 1892 to preserve and protect the park. The park is presently owned by 280 residents of the Larchmont Manor neighborhood, who generously allow public access to what otherwise would be a private domain.

Directions

Traveling on I-95/New England Thruway south of New Rochelle, take Exit 15 for Route 1/The Pelhams/New Rochelle. Head northeast on Route 1/ Main Street for <3.0 miles. Then turn right onto Beach Avenue and proceed southeast for 0.7 mile. When you come to the end of Beach Avenue, follow it as it now turns around and heads north as Park Avenue. You will see Manor Park to your right.

There doesn't appear to be any public parking near the beach. We would suggest, then, parking along either Kane Avenue or Larchmont Avenue (streets parallel to Beach Avenue), and walking back to the park.

Resources

Judith Doolin Spikes, *Larchmont, NY: People and Places* (Larchmont, NY: Fountain Square Books, 1991), 1.

Wikipedia, s.v. "Manor Park, Larchmont," last modified October 21, 2022, http://en.wikipedia.org/wiki/Manor_Park,_Larchmont.

"Larchmont Manor Park: When We Can't Get Away, Our Local Get-Away." https://fifiandhop.com/2015/05/29/when-we-cant-get-away-our-local-get-away.

"Geographic History – Larchmont Manor Park." https://larchmontmanorpark.org/geographic-history.html. This website not only talks about the geology of the park, but specifically mentions a rounded boulder called Dinosaur Egg.

"Larchmont Manor Park Society." https://larchmontmanorpark.org/the-park.html.

Washington Rock 44

Type of Formation: Rock Profile
WOW Factor: 3
Location: Mamaroneck (Westchester County)
Tenth Edition, NYS Atlas & Gazetteer: p. 111, B8
Earlier Edition, NYS Atlas & Gazetteer: p. 25, A5–6
Parking GPS Coordinates: *Delancey Road*: 40°56.686′N, 73°44.319′W; *Fenimore Road*: 40°56.746′N, 73°44.189′W
Washington Rock GPS Coordinates: 40°56.620′N, 73°44.312′W
Accessibility: Roadside
Degree of Difficulty: Easy

Description

Washington Rock is a medium-sized rock profile that either formed naturally or, as some believe, was intentionally sandblasted onto the rock face by a construction crew around 1890.

Some even believe the urban legend that the rock's facial outline was the handiwork of craftsmen from D. W. Griffith's film studio. Griffith is best remembered for his film *Birth of a Nation*. The only fly in the ointment is that Griffith didn't settle in Mamaroneck at Satan's Toe (a point of land) until 1919, eighteen years after the rock's uniqueness was noticed.

History

The Washington Rock property has exchanged hands a number of times. Originally, it was owned by the titleholders of the 1677 Henry Disbrow House, then later by proprietors of the Washington Arms Restaurant. It wasn't until 1901, however, that the rock's amazing likeness to George Washington's face was noticed by one of the townsfolk. We're not sure if an answer was ever given as to why it took so long for the rock's unusual facial likeness to be noticed. In any case, Washington Rock

20. Washington Rock. Antique postcard, public domain.

immediately became the rage, and postcards of it were reproduced in great numbers.

Then, as the years turned into decades, the rock gradually fell into obscurity until no one alive even knew of its exact location. In 2007, efforts to locate the lost rock paid off when it was rediscovered behind a thick growth of sumac and vines near roadside at the Liberty Montessori School.

Directions

From Mamaroneck (at the junction of Routes 1/Boston Post Road & 127/ Harrison Avenue), drive southwest on Route 1/Boston Post Road for 0.8 mile and turn left (south) onto Orienta Avenue. The rock will be immediately on your right, located on the slope adjacent to the Liberty Montessori School, directly across from Mamaroneck Harbor.

Nearby Delancey Avenue or Fenimore Road, both on the north side of Route 1, are places where you can temporarily park.

Resources

"From the Vault: Mamaroneck Rocks." https://larchmontloop.com/from-the-vault-mamaroneck-rocks/.

Stephenson Boulder 45

Stephenson Park

Type of Formation: Large Boulder
WOW Factor: 4–5
Location: New Rochelle (Westchester County)
Tenth Edition, NYS Atlas & Gazetteer: p. 111, BC7
Earlier Edition, NYS Atlas & Gazetteer: p. 25, A5
Parking GPS Coordinates: 40°55.057′N, 73°46.441′W
Stephenson Park GPS Coordinates: 40°55.106′N, 73°46.434′W
Stephenson Boulder GPS Coordinates: 40°55.121′N, 73°46.436′W
Accessibility: Near roadside
Degree of Difficulty: Easy

Description

We have not been able to find any specific background information about the Stephenson Boulder. It appears to be around 7–8 feet high and over 12 feet long, resting at the base of a tiny hillock.

History

Stephenson Park is a recreational area containing mainly an athletic field and playground.

Somehow, fittingly, the Rock Club and Climbing Gym [40°55.061′N, 73°46.594′W] is literally only a block away on 130 Rhodes Street.

Stephenson Park is named after Stephenson Boulevard, a divided highway that runs along the west side of the park.

Directions

From Main Street/Route 1 in New Rochelle, drive northwest on Stephenson Avenue for 0.2 mile to reach the southwest corner of the park. Turn right onto Lyons Place and either park along this short street or turn right onto Lispenard Avenue (a one-way street) and park.

Stephenson Park lies between Palmer Avenue and Lispenard Avenue; the boulder is almost exactly midway between the two streets. A walkway from Lispenard Avenue leads past the boulder.

Buses numbered 45 and 60 have routes that pass by the park.

The Bronx

The Bronx encompasses 57 square miles of land and water, and is named for Jonas Bronx (Bronck), a former Dutch landowner. It is the northernmost of New York City's five boroughs and the only part of New York City (except for a minor sliver) that is located on the mainland. It is defined by the Hudson River to the west, Harlem River to the southwest, East River to the southeast, Long Island Sound to the east, and Westchester County to the north.

Fort Number 8 Boulder

46

University Woods

Type of Formation: Medium-Sized Boulder
WOW Factor: 3–4
Location: University Heights (The Bronx)
Tenth Edition, NYS Atlas & Gazetteer: p. 110, CD5
Earlier Edition, NYS Atlas & Gazetteer: p. 24, A2
Parking GPS Coordinates: 40°51.550′N, 73°54.870′W (Essentially, park along Sedgwick Avenue)
University Woods Entrance GPS Coordinates: 40°51.579′N, 73°54.804′W
Fort Number 8 Boulder GPS Coordinates: Not determined
Accessibility: Variable depending where you parked
Degree of Difficulty: Moderately easy

21. Fort Number 8 Boulder. *Source*: Stephen Jenkins, *The Story of the Bronx: From the Purchase Made by the Dutch from the Indians in 1639 to the Present Day* (New York: G. P. Putnam's Sons, 1912), 346–347 insert.

Description

This medium-sized boulder marks the site of Fort Number 8, a small, four-pointed star British redoubt that commanded a view of the Harlem River from Fordham Heights.

History

Words have been chiseled into the boulder's surface, but a photograph that we have seen of the rock is too indistinct and the words too blurred to be legible. Most likely, the words convey an account of the site's history.

Fort Number 8 was built by the British in 1776 on one of the highest points in the Bronx. Its cannons roared in November of 1776 when Fort Washington, today known as Fort Knyphausen (see chapter on Fort Hill Park Boulder) was under attack.

The fort endured until 1782, when it was abandoned by the redcoats and then destroyed that same year.

In 1857, Gustav Schwab built a mansion on the site, which he and his wife, Eliza, lived in until their deaths. In the early 1900s, the house was acquired by New York University.

Number 8 Boulder is located in University Woods, aka Cedar Park—a section of the University Heights neighborhood between Sedgwick Avenue and Cedar Avenue.

University Heights was formerly known as Fordham Heights. The name of the community changed due to its association with the former New York University campus.

Directions

From the Major Deegan Expressway/I-87 following along the east side of the Harlem River, take Exit 9 for "West Fordham Road & University Heights Blvd." Go northeast on West Fordham Road for >0.2 mile and then turn right onto Sedgwick Avenue. Proceed southwest for over 0.3 mile and park along the street where you can. The park is directly to your right, located near the west side of Bronx Community College.

According to the website nycgovparks.org/parks/university-woods/history, "British Fort #8 occupied the crest of the ridge now known as University Woods."

We assume that the boulder should be easy to locate once you get there. It's quite possible, however, that for all your effort, the boulder's size may prove to be less than overwhelming.

Resources

Stephen Jenkins, *The Story of the Bronx: From the Purchase Made by the Dutch from the Indians in 1639 to the Present Day* (New York: G. P. Putnam's Sons, 1912), 346–347. A photograph of the boulder at Fort Number 8 is shown in an insert.

"Fort No. 8 NYC." www.fortwiki.com/Fort_No._8_-_NYC.

"University Woods." https://www.nycgovparks.org/parks/university-woods/history.

Indian Rock 47

Crotona Park

Type of Formation: Large Boulder
WOW Factor: 6
Location: Claremont (The Bronx)
Tenth Edition, NYS Atlas & Gazetteer: p. 111, CD6
Earlier Edition, NYS Atlas & Gazetteer: p. 25, B4–5
Crotona Park GPS Coordinates: 40°50.284′N, 73°53.757′W
Indian Rock GPS Coordinates: 40°50.257′N, 73°53.715′W
Accessibility: 0.1-mile walk from Crotona Avenue
Degree of Difficulty: Easy

Description

In *South Bronx Rising: The Rise, Fall, and Resurrection of an American City*, Jill Jonnes writes, "Indian Rock, a tall boulder, was a favorite destination for clambering and adventure." It still is today.

According to John McNamara, in *History in Asphalt: The Origin of Bronx Street and Place Names*, the rock measures 6 feet by 8 feet with a height of 10 feet and "has four deeply-cut steps on its north face. These steps are evenly spaced and resemble stirrups."

History

The 127-acre Crotona Park is named after the Greek colony of Croton. The land was acquired by the city in 1888. By 1914, the perimeter of 3.3-acre Indian Lake was fortified with concrete walls, and pathways were established around the lake and through sections of the park.

Like with many bodies of water before modern refrigeration, ice was harvested from Indian Lake during the winter for summer cooling.

Indian Rock is believed to have been named in the late 1800s by local youths who, engaging in play by the rock, would conjure up images of Native Americans and great council meetings.

The rock has also been used for political gatherings. In "The History of Indian Lake—Crotona Park," Harvey Lubar writes that "Socialist rallies were occasionally held at the lake and their fiery speakers tried to excite the crowds from the top of Indian Rock."

The land was originally part of the extensive, 140-acre Alexander Bathgate farm during the nineteenth century.

Directions

There really is no easy way to get to Crotona Park, and you may need to plot your own course.

However, we did find one relatively easy route. If you are traveling east on the busy Cross Bronx Expressway/I-95, take Exit 2B for Webster Avenue. When you come to Webster Avenue, turn right and drive southwest for 0.3 mile. At Claremont Parkway, turn left and proceed southeast for 0.3 mile. This highway takes you right into the park, where limited roadside parking is available in places.

A second option is to turn left onto Fulton Avenue (a one-way street) and head around the 2-mile perimeter of the park, from Crotona Park North to Crotona Park East to Crotona Park South, all one-way, so that you will be continually traveling clockwise. There are many places to park around the park.

Indian Rock is situated on a small hill overlooking the southwest end of Indian Lake, 75 feet uphill from the walkway that circles around the lake.

Resources

Jill Jonnes, *South Bronx Rising: The Rise, Fall, and Resurrection of an American City* (New York: Fordham University Press, 2002), 212.

Harvey Lubar, "The History of Indian Lake," *Bronx County Historical Society* 22, no. 2 (Fall 1985): 54. "A large boulder on the southwest corner of the lake was named 'Indian Rock' as the boys named the rock the chief's seat.

John McNamara, *History in Asphalt: The Origin of Bronx Street and Place Names* (Harrison, NY: Harbor Hill Books, 1978), 381.

"Bronx Diary - Crotona Park." https://www.bronxboard.com/diary/diary.php?f = Crotona%20Park.

Harry T. Cook, *The Borough of the Bronx, 1639–1913* (New York: Author, 1913), 68.

Cat Rock Cave (Historic) 48

Type of Formation: Cave
WOW Factor: Unknown
Location: Rochelle Heights (Westchester County)
Tenth Edition, NYS Atlas & Gazetteer: p. 111, C7
Earlier Edition, NYS Atlas & Gazetteer: p. 25, A5
Junction of Rockland Place and Lemke Place GPS Coordinates: 40°55.469′N, 73°46.418′W
Cat Rock Cave GPS Coordinates: 40°55.510′N, 73°46.433′W (guesstimate)
Accessibility: The cave is undoubtedly on private land that was significantly altered by developers.

Description

According to Morgan Secord in his article, "Cat Rock Cave," Cat Rock was a true cave, and not just a rock-shelter. It originally extended into a rocky ridge. However, when the land around Rockland Place (the street) underwent further development, a considerable volume of the ridge was sliced off.

At one point, the cave was visited by Mark W. Harrington, engaged in archaeological research for the Museum of the American Indian, who found stone implements, suggesting that the cave had been used by Native Americans.

In the article entitled "Elastic Cave," the author writes, "Legends about the cave are probably the basis for tales of an underground tunnel between Port Chester and New Rochelle."

In the April 1955 issue of *The Westchester Historian: Quarterly of the Westchester County Historical Society*, "A field trip to the area revealed a high shattered rock mass facing Rockland Place about three hundred feet west of Lemke Place. Its natural, shattered appearance had been made even more so by other blasting which destroyed a small natural cave during the improvement of Rockland Place some years ago."

It would seem that nothing is left of the cave today.

But wait! Morgan Secord in his article on Cat Rock Cave goes on to say that "originally, there existed in front of Cat Rock Cave a thickly strewn area of dozens of large boulders, dropped during the ice age covering over half an acre. These stones varied in size, many larger than half a barrel." Perhaps there is still something there to see after all.

Directions

From Rochelle Heights (junction of 5th Avenue & North Avenue), go northeast on 5th Avenue for 0.8 mile. Turn right onto Rockland Place (a one-way street until Pierce Street) and head southeast for 0.3 mile to reach the junction with Lemke Place.

The area is only 0.1 mile north of the New England Thruway/I-95.

Morgan Secord places the location of the cave "to the east or rear of Rochelle Heights, a subdivision adjoining the east side of Rockland Place, north of Lemke Place." The GPS coordinates given above are for the junction of Rockland Place and Lemke Place. We suspect the cave is or was only several hundred feet from this spot, where there is a wooded area between houses.

Resources

Morgan Secord, "Cat Rock Cave," *Westchester Historian: Quarterly of the Westchester County Historical Society* 41, no. 4 (Autumn 1965): 68.

"Elastic Cave." *Westchester Historian: Quarterly of the Westchester County Historical Society* 31, no. 2 (April 1955): 54. A field trip to Cat Rock Cave is described.

Herbert B. Nichols, *Historic New Rochelle* (New Rochelle, NY: Board of Education, 1938), 170.

Split Rock and Lincoln Rock 49

Bronx Park: New York Botanical Garden

Type of Formation: Split Rock
WOW Factor: 3
Location: Bronx
Tenth Edition, NYS Atlas & Gazetteer: p. 111, C6
Earlier Edition, NYS Atlas & Gazetteer: p. 25, AB4–5
Parking GPS Coordinates: *Main*: 40°51.708′N, 73°52.844′W;
Parking Area A: 40°51.627′N, 73°52.887′W
Destination GPS Coordinates: *Split Rock*: 40°51.833′N, 73°52.721′W;
Lincoln Rock: Unknown
Fee: Admission charged
Days & Hours: Check website for specific details
Accessibility: *Split Rock*: 0.2-mile walk
Degree of Difficulty: Easy
Additional Information: New York Botanical Garden,
2900 Southern Boulevard, Bronx, NY 10458
Botanical Garden Map: mappery.com/map-of/The-New-York-Botanical-Garden-Map

---

Description

The New York Botanical Garden Split Rock is a 6-foot-high, 8-foot-long boulder made of Fordham gneiss that was broken into two uneven pieces by the roots of a long-vanished tree. The larger half is about two-thirds of the rock's total length.

A second point of interest is Lincoln Rock. In *History in Asphalt: The Origin of Bronx Street and Place Names*, John McNamara describes the rock as a high point at the gardens west of the Bronx River where "a likeness of

22. Bronx Park Split Rock. Photograph by the author.

the president [was] cut into the rock sometime in the 1940s." We have not been able to find any current information about this profile unfortunately.

History

The 250-acre botanical garden was created in 1891 and the 2.5-acre rock garden in the 1930s.

The Bronx River—New York City's only freshwater stream—runs through the park. More is said about the river in the next chapter.

Directions

From the Bronx River Parkway, take Exit 8 and proceed southwest on Southern Boulevard/Dr. Theodore Kazimiroff Boulevard for >1.0 mile. Get off the exit for the botanical garden and park in the main area or area A.

You may also wish to consult the New York Botanical Garden website (www.nybg.org) for more specific directions depending upon your starting point.

Split Rock is located inside the botanical garden not far from the main entrance. Once you pass through the entrance, turn right at the circular pond and then, after 0.1 mile, left into the entrance to the Native Plant Garden. Walk to the left of the small pond to quickly reach the historic rock.

We don't know exactly where Lincoln Rock is or whether it still exists, but John McNamara's description suggests that you should look for a high point in the gardens.

Resources

"New York Botanical Garden." https://www.nybg.org. Main site for the New York Botanical Garden.

"Split Rock at the New York Botanical Gardens." https://worleygig.com/2020/01/09/split-rock-at-the-new-york-botanical-garden/.

Allen Rokach, "History Underfoot: A Short Geological History of the Bronx," *Bronx County Historical Society Journal* 11, no. 2 (Fall 1974): 71–80.

Randall Comfort, comp., *History of Bronx Borough: City of New York* (New York: North Side News Press, 1906), 2. Mention is made of Indian Well and Bear's Den.

John McNamara, *History in Asphalt: The Origin of Bronx Street and Place Names* (Bronx, NY: Bronx County Historical Society, 1984), 398.

Andrew S. Dolkart, *Guide to New York City Landmarks* (New York: John Wiley & Sons, 1998), 260.

Bronx River Gorge Potholes 50

Bronx Park: New York Botanical Garden

Type of Formation: Pothole
WOW Factor: 4
Location: Bronx
Tenth Edition, NYS Atlas & Gazetteer: p. 111, C6
Earlier Edition, NYS Atlas & Gazetteer: p. 25, AB4–5
Parking GPS Coordinates: 40°51.709′N, 73°52.846′W
Destination GPS Coordinates: Not determined
Fee: Admission charged
Days & Hours: Check website for specific details
Accessibility: Not determined
Degree of Difficulty: Moderate
Additional information: New York Botanical Garden, 2900 Southern Boulevard, Bronx, NY 10458
Botanical Garden Map: mappery.com/map-of/The-New-York-Botanical-Garden-Map

---

Description

A number of potholes have formed in the bedrock of the Bronx River Gorge. In *The Borough of the Bronx, 1639–1913*, Harry T. Cook writes, "The Bronx River runs directly thru part of the park from north to south varying in width from 50 to 400 feet." This is plenty of space where potholes can form.

There are other geological features also worthy of note that are located, we believe, in the area of the Bronx River. Indian Well, aka Indian Bath, is described as "a rocky basin perhaps used by the red men as a place to grind their corn that is located inside a cliff." We suspect that this is one of the potholes. In addition, there is Bear's Den, "a romantic spot where the rocks were piled perpendicularly by some immense force, between them being a natural cave," in other words, a rock-shelter.

History

Geologist believe that during the last Ice Age, the Bronx River emptied into the Hudson River before a huge ice mass blocked the river's flow west, forcing it to cut its present channel into the East River and Long Island Sound. It is for this reason that the gorge is so deeply cut in the area by the New York Botanical Garden. Native Americans called the river Aquehung, or "River of High Bluffs," due to its 75-foot-high walls. Early on, it served as a major border between the Wappinger and Siwanoy tribes.

The gorge starts half a mile or so downriver from the Williams Bridge and ends before a rocky waterfall at Bronxdale.

Like many industrialized rivers, the Bronx River had become an open sewer by the early 1900s. Fortunately, during the years that followed, a number of advocacy groups began working to clean up the river, culminating with the Bronx River Alliance in 2001.

Directions

See previous chapter for directions.

The Bronx River is 0.3–0.4 linear miles east of the parking area. The best route to take to get to the river seems to be to follow the Forest Trail, which approaches the river in at least two spots. We suspect that there is an informal path along the river.

Other entry points to the river are from the Hester Bridge or Snuff Mill Bridge, both within the park. The Snuff Mill Bridge was named for a mill built in 1840 that ground tobacco into snuff. It was acquired by the botanical garden in 1915. The mill now houses a café.

In addition, it may be possible to explore the Bronx River by kayak or canoe, putting in at Shoelace Park [40°53.359'N, 73°51.891'W off East 233rd Street] and debarking at Hunts Point Riverside Park [40°49.063'N, 73°52.901'W off Lafayette Avenue]. The journey will require three short portages, and you are not permitted to enter the grounds of either the botanical garden or Bronx Park from this river journey. For more details, consult Kevin Stiegelmaier's 2009 book *Canoeing and Kayaking New York*.

We have no idea exactly where the potholes or the Bear's Den are along the river, but that's what makes the exploration part of this book so much fun. Take heart. You will be in good company, for Edgar Allan Poe and

Joseph Rodman Drake (an early nineteenth-century poet best known for "The Culprit Fay") once walked along the same pathways and riverbed.

Resources

Lloyd Ultan, *The Northern Borough: A History of the Bronx* (Bronx, NY: Bronx Historical Society, 2009), 3. "Many [potholes] can be found in the Bronx River Gorge at the northeast end of the forest in New York Botanical Garden."

"Bronx River Parkway." https://www.nycgovparks.org/parks/X004/highlights/11583.

C. R. Roseberry, *From Niagara to Montauk: The Scenic Pleasures of New York State* (Albany, NY: State University of New York Press, 1982), 278–279.

"Bronx River Guided Folding Booklet." https://www.nybg.org/content/uploads/2017/03/BronxRiverSelf-GuidedVisitTeacherGuide.pdf.

Christopher J. Schuberth, *The Geology of New York City and Environs* (Garden City, NY: Natural History Press, 1968), 88. The author mentions that the river's course was altered when it eroded "a new, narrow gorge along this fault through the mica schist, thereby abandoning the lower part of its original way." Schuberth believes this occurred some 20–30 million years ago.

Harry T. Cook, *The Borough of the Bronx. 1639–1913* (New York: printed by the author, 1913), 67.

Kevin Stiegelmaier, *Canoeing and Kayaking New York* (Birmingham, AL: Menasha Ridge Press, 2009), 176–180.

Herbert B. Nichols, *Historic New Rochelle* (New Rochelle, NY: Board of Education, 1938), 112. "In Bronx Park a big stone can still be seen in the bottom of a large pothole."

Rocking Stone 51

Bronx Park: Bronx Zoo

Type of Formation: Rocking Stone
WOW Factor: 4
Location: Bronx
Tenth Edition, NYS Atlas & Gazetteer: p. 111, C6
Earlier Edition, NYS Atlas & Gazetteer: p. 25, AB4–5
Parking GPS Coordinates: 40°51.179′N, 73°52.373′W (Bronx River parking lot)
Rocking Stone GPS Coordinates: 40°50.876′N, 73°52.674′W (guesstimate)
Fee: Admission charged
Days & Hours: Check website, bronxzoo.com, for specific details
Accessibility: 0.4-mile walk
Degree of Difficulty: Easy
Additional Information: Bronx Zoo, 2300 Southern Boulevard, Bronx, NY 10460
Bronx Zoo Map: bronxzoo.com/map.

Description

The Bronx Zoo Rocking Stone is a pink, granite rock that measures 7 feet high by 10 feet wide. It rests on top of a bedrock outcrop. Some estimate its weight to be up to 30 tons. John McNamara, in *McNamara's Old Bronx*, describes it as "a rough cube of pinkish granite . . . that could be tilted a scant 2 inches. So much for the boulder moving like a rocking horse."

The bedrock upon which the boulder rests reveals very defined groove marks that were caused by stones being rubbed together as they were pushed forward by glaciers.

The rock has been a major attraction at the zoo since 1895 when it was featured on postcards and frequently used as a backdrop for family photographs.

23. Bronx Zoo Rocking Stone. *Source*: Stephen Jenkins, *The Story of the Bronx: From the Purchase Made by the Dutch from the Indians in 1639 to the Present Day* (New York: G. P. Putnam's Sons, 1912), 310–311 insert.

Because visitors were always trying to dislodge the boulder by rocking it, the zoo officials decided to play it safe, eventually shoring up the boulder's base to prevent it from rocking at all.

History

According to legend, a team of 24 oxen once tried to dislodge the rock without success. That's a lot of power. Whether this is true or not is within the realm of debate.

Like many large boulders or rocks with unusual properties, the Rocking Stone served as a convenient landmark for colonial surveyors, thereby fixing the northern boundary that made up the original 12 West Farms.

The Rocking Stone Restaurant, which often shows up in the background on postcards featuring the Rocking Stone, operated until 1942.

In his book, *History of Bronx Borough. City of New York*, Randall Comfort includes a brief poem about the rock:

> A rock, chance poised and balanced lay,
> So that a stripling arm might sway,
> A mass no host could raise.
> In nature's rage at random thrown

Yet trembling like the Druid's stone
On its precious base.

The Bronx Zoo, aka New York Zoological Park, encompasses 265 acres of land and is partially bisected by the Bronx River. It opened in 1898.

Directions

Driving along the Bronx River Parkway, take Exit 6, which leads immediately to the east side (Gate B) parking area for the Bronx Zoo.

You may also wish to consult the Bronx Zoo website (bronxzoo.com) for more specific directions depending upon your starting point.

The boulder is located next to the former House (World) of Darkness that opened in 1969 and closed in 2009. The House of Darkness was built on the site of the former Rocking Stone Restaurant.

Using Google Earth, we were able to locate a building called the "Nocturnal House," 0.4-linear miles southwest of the parking area and saw what looked like a large boulder on the walkway leading to it. Hopefully, what our GPS coordinates indicate is the actual site of the Rocking Stone.

Resources

John Scheier, *New York City Zoos and Aquarium: Images of America* (Charleston, SC: Arcadia Publishing, 2005). A photograph of the rock is shown on page 27.

"The Bowery Boys. New York City History." https://www.boweryboyshistory.com/2010/04/bronx-zoo-tale-behind-nycs-biggest.html. This website shows a postcard image of the Rocking Stone.

Randall Comfort, comp., *History of Bronx Borough: City of New York* (New York: North Side News Press, 1906). On page 2 is a photograph of the Rocking Stone, with tiny girls posed at each end as if trying to rock it back and forth.

John McNamara, *McNamara's Old Bronx* (Bronx, NY: Bronx County Historical Society, 1989), 102.

Diana Farkas, "The Zoological Park, Bronx, N.Y." *Bronx County Historical Society Journal* 11, no. 1 (Spring 1974). On page 2 is a 1908 photograph of the Rocking Stone.

Stephen Jenkins, *The Story of the Bronx: From the Purchase made by the Dutch from the Indians in 1639 to the Present Day* (New York: G. P. Putnam's Sons, 1912), 308. "The 'Rocking-stone' is an immense boulder weighing several tons, left here by some melting glacier, whose course is plainly marked by the scratche [*sic*] on the exposed rock surface. The boulder is so nicely balanced that a slight force will set it rocking."

Lloyd Ultan, *The Northern Borough: A History of the Bronx* (Bronx, NY: Bronx Historical Society, 2009), 3. "When it dropped, it was delicately perched atop a Fordham gneiss outcropping in such a way that it could be tipped back and forward, giving it its name."

Harry T. Cook, *The Borough of the Bronx. 1639–1913* (New York: printed by the author, 1913), 2. Cook calls it "a colossal cube of pinkish granite." On page 66, he also calls it a "100-ton Rocking Stone."

Allen Rokach, "History Underfoot: A Short Geologic History of the Bronx," *Yonkers Historical Bulletin* 11, no. 2 (Fall 1974): 71–80.

Black Rock 52

Soundview Park

Type of Formation: Large Rock
WOW Factor: 5
Location: Bronx
Tenth Edition, NYS Atlas & Gazetteer: p. 111, D6
Earlier Edition, NYS Atlas & Gazetteer: p. 25, B4–5
Parking GPS Coordinates: 40°49.196′N, 73°52.403′W
Soundview Park GPS Coordinates: 40°48.728′N, 73°51.936′W
Black Rock GPS Coordinates: Not determined
Accessibility: Could be up to a 0.5-mile walk depending on exactly where the boulder is located
Degree of Difficulty: Moderately easy
Park Map: nycgovparks.org/parks/soundview-park/map

Description

In *History in Asphalt: The Origin of Bronx Street and Place Names*, John McNamara writes, "This great boulder imbedded in the salt marshes near the junction of Ludlow's Creek and the Bronx River was thought by the early inhabitants of Clason Point to be a meteorite."

Randall Comfort, in *History of Bronx Borough: City of New York*, writes that Black Rock is a great boulder that lies "partially imbedded in the salt marshes to the south of Westchester Turnpike, not far from Pugsley's Causeway."

In the 1830s, Black Rock became part of Ludlow's Black Rock Farm on Clason Point.

The boulder is said to be formed of gneiss, a coarse-grained, imperfectly layered metamorphic rock.

History

Ludlow Creek, mentioned by McNamara, doesn't appear on current maps except for Ludlow Creek (not the same one) in Suffolk County.

It's possible that Pugsley's Causeway, which is not identified by name today, may be the causeway at the southeast end of Soundview Park. If that is so, then one might wonder why MapQuest places Pugsley Creek on the east side of Clason Point. It can get confusing when you try to reconcile the past with the present, for land features have changed significantly over the last couple of hundred years. Complicating matters further is that there is also a Blackrock Playground [40°49.709'N, 73°51.477'W] bounded by Watson Avenue, Blackrock Avenue, and Pugsley Avenue—a site that may have played a role in Black Rock's history.

Fortunately, we know exactly where the boulder is today. According to the New York City Department of Parks and Recreation's excellent website nycgovparks.org/parks/black-rock-playground/history, after being "mistaken for a meteorite by early settlers, the boulder was moved to Soundview Park, where it can be seen today."

The 205-acre Soundview Park, the so-called "gateway to the Bronx River," opened in 1937. At that time, the entire area was composed of marshland, with three streams running through it. Landfill operations lasted forty years, bringing the height of the shoreline up to 30 feet above its starting level. To be sure, the park looks nothing like it did in the 1930s.

Directions

Approaching the south terminus of the Bronx River Parkway, get off at the last exit, which is for "Story Avenue & Sound View Park." Continue straight ahead onto Morrison Avenue and continue south for >0.2 mile. Then bear right (west) onto Lafayette Avenue and park on either side of the road. The entrance to Soundview Park, which you just passed by, is at the corner of Morrison Avenue and Lafayette Avenue.

We have not been able to determine where the large rock is located in Soundview Park, but we suspect it's on full display, since the park obviously went to some trouble to preserve it.

Resources

John McNamara, *History in Asphalt: The Origin of Bronx Street and Place Names* (Harrison, NY: Harbor Hill Books, 1978), 287.

"Soundview Park." https://www.nycgovparks.org/parks/soundview-park/history.

Randall Comfort, comp., *History of Bronx Borough: City of New York* (New York: North Side News Press, 1906), 2.
Harry T. Cook, *The Borough of the Bronx, 1639–1913* (New York: printed by the author, 1913). Cook mentions that Black Rock is on "Westchester Avenue just above the old Watson Estate and the Westchester Golf Club." One should bear in mind that this account was written over 100 years ago.
Stephen Jenkins, *The Story of the Bronx: From the Purchase Made by the Dutch from the Indians in 1639 to the Present Day* (New York: G. P. Putnam's Sons, 1912), 402.

Inwood Hill Park Rock Formations 53

Inwood Hill Park

Type of Formation: Pothole; Rock-Shelter
WOW Factor: 5–6
Location: Inwood (The Bronx)
Tenth Edition, NYS Atlas & Gazetteer: p. 110, C5
Earlier Edition, NYS Atlas & Gazetteer: p. 24, A2
Parking GPS Coordinates: Park along side streets where spaces are available
Destination GPS Coordinates: *Potholes*: 40°52.225′N, 73°55.580′W & 40°52.287′N, 73°55.525′W; *Rock-Shelter*: 40°52.390′N, 73°55.502′W; *Shorakkopoch Rock*: 40°52.395′N, 73°55.420′W; *Indian Caves*: 40°52.378′N, 73°55.485′W (Google Earth)
Accessibility: 0.2–0.9-mile walk
Degree of Difficulty: Moderately easy
Trail Map: nycgovparks.org/pagefiles/82/Inwood-Hill-Park-map_2014.pdf

---

Description

A number of potholes can be found at the park. Of these, the Inwood Hill Park Pothole is the largest, and allegedly the largest of its kind in New York City. The pothole's overlying mica schist has been worn away to expose the underlying Inwood marble.

In *The Geology of New York City and Environs*, Christopher J. Schuberth mentions that these series of subglacial potholes were formed by an "eddy in water of a stream flowing beneath the melting ice of the Wisconsin Glacier." They are found in an area of the park called The Clove, and were first "discovered" in 1931 by Patrick Coghlan, an Inwood resident.

We have seen estimates of the potholes' sizes being as large as 5 feet deep and 3.5 to 8 feet in diameter.

The rock-shelters are formed from configurations of slabs of Manhattan schist that were literally torn out of the bedrock by glaciers. The main rock-shelter was "discovered" by Alexander Chenoweth in 1890 who opened up the cave by digging through mounds of earth that had washed into the mouth of the cave over the previous two centuries.

One of the formations, called the Inwood Hill Tunnels, was bricked over in the 1920s out of concerns for public safety from illicit activities taking place (this was similarly done in the Rambles section of Central Park).

Reginald Pelham Bolton, in *Washington Heights, Manhattan: Its Eventful Past*, mentions an "overhanging rock-shelter in which were found layers of ashes that had formed in the household fires."

Shorakkopoch Rock is a small boulder that marks the general area where Peter Minuit bought Manhattan Island from Native Americans for what amounted to 60 guilders (the equivalent of $24 today). The concrete circle around the boulder represents the circumference of a great tulip tree that once occupied this spot.

In his book Bolton also mentions Spouting Spring, where "above this ancient outlet, a great slab of rock is exposed, in which are formed by a freak of nature, aided perhaps by some human labor, three depressions which take the form of eyes and a beak or nose." Bolton goes on to speculate that the name of the Spuyten Duyvil Creek may have originated from this spring, since a 1672 document describes it as "Spuyten Duyvil alias the Fresh Spring."

History

Inwood Hill, formerly called Cock Hill and, before that, New Haarlem in 1677, was earlier the site of a Native American village called Shorakapkok (meaning "as far as the sitting-down place"). The hill, sheltered from icy winds that made Manhattan's rocky ridge less hospitable, proved to be an ideal spot for a village.

Inwood Hill Park officially opened in 1917, but land purchases continued to increase the size of the park until 1941. It contains 196 acres.

The Hudson River Bike Trail and secondary hiking trails provide ready access to the park's interior. The trails were constructed in the 1930s by the Work Projects Administration.

Directions

Inwood Hill Park is located at the northwest corner of Manhattan near Route 9, north of Fort Tryon Park.

From I-87/Major Deegan Expressway, take Exit 9 for West Fordham Road and University Heights Boulevard. Head northwest (away from University Heights and West Fordham Road), crossing over University Heights Bridge and continuing northwest on West 207th Street for 0.6 mile until you come to Inwood Hill Park.

Parking, of course, is always problematic where no designated parking areas exist. Nearby streets, such as Dyckman Street or Payson Avenue (a one-way street with parking on both sides), may afford spaces to park.

The main geological features are shown on the Inwood Hills trail map. To reach the glacial potholes (listed on the map as site #10), follow the ravine down the east side of the ridge. The Indian Rock Shelters (#9) are encountered further down, through a grove of spicebush. Shorakkopoch Rock (at #7) is just slightly west of the rock shelter formations.

We have no idea where Spouting Spring is located, whether it still exists, or whether it is even located within the borders of the park.

Resources

Christopher J. Schuberth, *The Geology of New York City and Environs* (Garden City, NY Natural History Press, 1968), 224.

"Inwood Hills Park Hiking Trails." https://www.nycgovparks.org/park-features/hiking/inwood-hill-park.

"Glacial Pothole in the City." https://www.geocaching.com/geocache/GC2RWPQ_glacial-pothole-in-the-city.

"Inwood Hill Park." https://www.nycgovparks.org/parks/inwood-hill-park.

"Glacial Potholes of Inwood Hill Park" https://myinwood.net/glacial-potholes-of-inwood-hill-park.

Lisa Montanarelli, *New York City Curiosities: Quirky Characters, Roadside Oddities and Other Different Stuff* (Guilford, CT: Morris Book Publishing, 2011), 54–55. A photograph of one of the boulders is shown on page 54. On page 40 is a photograph taken while looking out from inside the shelter cave.

"Shorakkopoch Rock – New York, New York – Atlas Obscura." https://www.atlasobscura.com/places/shorakkopoch.

"Inwood Hill Park Hiking Trails." https://nycgovparks.org/park-features/hiking/inwood-hill-park. This website also includes a map that shows the relative positions of the glacial potholes and rocks.

Reginald Pelham Bolton, *Washington Heights, Manhattan: Its Eventful Past* (New York: Dyckman Institute, 1924), 13. On page 15 is a photograph of one of the

park's rock-shelters. Bolton devotes an entire chapter to the park, pages 170–173, and to Tubby Hook, pages 172–178.
Reginald Pelham Bolton, *Indian Life of Long Ago in the City of New York* (New York: Bolton Books, 1924). Between pages 123 and 125 is an insert containing an illustration of how Native Americans used the rock overhangs for shelter by draping poles in front of the overhangs, and then closing them off using skins or sheets of bark.
C. R. Roseberry, *From Niagara to Montauk: The Scenic Pleasures of New York State* (Albany, NY: State University of New York Press, 1982), 275. "Huge, fallen wedges of rock from the Inwood Heights form a disorderly cluster locally known as the Indian Caves."
Sanna Feirstein, *Naming New York Manhattan: Places and How They Got Their Names* (New York: New York University Press, 2001), 176, 178.
Lee Ann Levinson, *East Side, West Side: A Guide to New York City Parks in All Five Boroughs* (Darien, CT: Two Bytes, 1997), 60. "Still today, caves in the park have Indian relics."
Charles Merguerian and Charles A. Baskerville, "Geology of Manhattan Island and the Bronx, New York City, New York," in *Northeastern Section of the Geological Society of America*, ed. David C. Roy, vol. 5, *Centennial Field Guide* (Boulder, CO: Geological Society of America, 1987), 137-149.
"The Indian Caves of Inwood Hill Park." https://myinwood.net/the-indian-caves-of-inwood-hill-park.
Raymond H. Torrey, Frank Place Jr., and Robert L. Dickinson, *New York Walk Book*, 3rd ed. (New York: American Geographical Society, 1951), 8. "From the top of Inwood Hill, follow the ravine down the east side of the ridge, by the great potholes in the exposed rock."

Fort Tryon Rocks 54

Fort Tryon Park

Type of Formation: Large Boulder
WOW Factor: 4–5
Location: Fort Tryon (The Bronx)
Tenth Edition, NYS Atlas & Gazetteer: p. 110, C5
Earlier Edition, NYS Atlas & Gazetteer: p. 24, B4
Parking GPS Coordinates (Cloisters Museum & Gardens): 40°51.903′N, 73°55.849′W
Destination GPS Coordinates: *Fort Tryon*: 40°51.758′N, 73°55.993′W; *Zombie Rock*: 40°51.931′N, 73°55.803′W; *Glacial Boulder*: 40°51.746′N, 73°56.000′W. There are a number of other boulders as well.
Accessibility: All are within a 0.5-mile walk
Degree of Difficulty: Moderately easy
Park Map: nycgovparks.org/parks/M029/map/ft-tryon-park-map.pdf

---

Description

It is entirely possible that many of the rocks in the park are rock buttresses or cliff-faces, for the word "rock" is frequently used indiscriminately to refer to either a boulder or a rock outcrop.

Boulderers have given the rocks such quaint names as Zombie Rock, Bulging Boulder, Ivy Rock, Guns, Cars Boulder, Life is Beautiful Boulder, Sherman Boulders, and Pathway Boulder (obviously next to the pathway). Some of these, by name, certainly sound like boulders.

One notable fact is that at a height of 267 feet, Fort Tryon Park is the highest point of land in Manhattan.

History

Fort Tryon Park is named for a Revolutionary War redoubt constructed by the Americans that fell to British forces in 1776. That redoubt, in turn, was named for William Tryon, a Loyalist and the last colonial governor of New York.

The property was later owned by Cornelius Kingsley Garrison Billings, a wealthy industrialist tycoon, from 1907 to 1916.

In 1917, John D. Rockefeller Jr. began acquiring land that would eventually become the park. In 1931, Rockefeller gifted the property to the public and then, from 1931 to 1935, the Olmsted Brothers firm, led by Frederick Law Olmsted Jr., created the 67-acre park that we see today.

The large boulders are believed to have been carried down by the Wisconsin Glacier from the Palisades. The bedrock that forms the outcroppings is made out of Manhattan schist.

The park offers commanding views of the New Jersey Palisades, Hudson River, and George Washington Bridge.

The area, known by the Lenape as Chquaesgeck, was first called Lange Bergh (Long Hill) by Dutch settlers.

Fort Tryon Park was added to the National Register of Historic Places in 1978 and to the New York City Scenic Landmark Registry in 1983. It serves as a memorial to Revolutionary War soldiers from Maryland and Virginia who fought against Hessian troops in 1776.

Directions

Fort Tryon Park lies just south of Inwood Hill Park.

From I-87/Major Deegan Expressway, take Exit 9 for West Fordham Road & University Heights Boulevard. Head northwest (away from University Heights and West Fordham Road), crossing over the University Heights Bridge and continuing on West 207th Street for >0.4 mile. Turn left onto Broadway and head southeast for 1.4 miles. When you come to West 181st Street, turn right and proceed west for 0.1 mile. Then turn right onto Fort Washington Avenue and head north for 0.6 mile until you come to Margaret Corbin Circle, the park's entrance.

The Fort Tryon Park Trust website, forttryonparktrust.org/visit-and-park-map/getting-here, also provides directions from a variety of approaches that you may find helpful.

Limited free parking is available at the New Leaf Restaurant and the Metropolitan Cloisters.

The vehicle entrance to the park is from Margaret Corbin Circle at the intersection of Fort Washington Avenue and Cabrini Boulevard.

Zombie Rock is located near the north end of the park, northeast of the Cloisters Museum and Gardens.

Many of the rocks are located near the bike path.

Resources

"Zombie Rock – Fort Tryon Park." https://rockclimbing.com/photos/Topo/Zombie_Rock_119944.html.

"Fort Tryon Park, Bouldering." https://www.thecrag.com/climbing/united-states/new-york-city/area/12665821.

Sanna Feirstein, *Naming New York Manhattan Places and How They Got Their Names* (New York: New York University Press, 2001), 171.

Lee Ann Levinson, *East Side, West Side: A Guide to New York City Parks in All Five Boroughs* (Darien, CT: Two Bytes, 1997), 43–45.

Christopher J. Schuberth, *The Geology of New York City and Environs* (Garden City, NY: Natural History Press, 1968), 220. A map shows the location of a glacial erratic in the park that the author believes was carried there from the Palisades.

Reginald Pelham Bolton, *Washington Heights, Manhattan: Its Eventful Past* (New York: Dyckman Institute, 1924), 158–162.

Jeffrey Perls, *Paths along the Hudson: A Guide to Walking and Biking* (New Brunswick, NJ: Rutgers University Press, 2001), 145–147.

Richmond (Echo) Park Rocks 55

Richmond (Echo) Park

Type of Formation: Medium-Sized Rock
WOW Factor: 4
Location: Mount Hope (The Bronx)
Tenth Edition, NYS Atlas & Gazetteer: p. 111, C6
Earlier Edition, NYS Atlas & Gazetteer: p. 25, B4–5
Parking GPS Coordinates: 40°50.955′N, 73°54.058′W (parking along Valentine Avenue)
Richmond (Echo) Park GPS Coordinates: 40°50.988′N, 73°54.069′W
Accessibility: <0.1-mile walk
Degree of Difficulty: Easy

Description

Some interesting rocks can be found in the park, particularly huge upthrusts of bedrock. Hopefully, a boulder or two will make its presence known.

The park is <0.2 mile long and 0.1 mile across at its widest.

History

Richmond Park, aka Julius Richmond Park, was a favorite outing for early New Yorkers. Visitors would come to the park, cup their hands in front of their mouths, and shout loudly just to hear their voices echo between the park's two rocky ridges. It was for this reason that the city named it Echo Park when they acquired the property in 1888. Politics won out in 1973, however, and the name changed to Richmond (Echo) Park in honor of a deceased civic leader, Julius J. Richmond, who was chairman of the Twin Parks Association and the Urban Action Task Force, as well as assistant administrator of the city's finance committee.

Richmond Park is delineated by Valentine Avenue to the east, East Tremont Avenue to the south, East Burnside Avenue to the north, and partly by Ryer Avenue to the west.

Directions

From the Bronx River Parkway, take Exit 7 and head northwest on Route 1/East Fordham Road for >1.0 mile. Turn left onto Webster Avenue, continuing on Route 1, and head southwest for 0.5 mile. Turn right onto East 182nd Street (a one-way street) and proceed west for <0.2 mile. When you come to Valentine Avenue, turn left and drive southwest for 0.3 mile. Park along Valentine Avenue. Richmond (Echo) Park is directly to your right.

Resources

John McNamara, *History in Asphalt: The Origin of Bronx Street Names and Place Names* (Harrison, NY: Harbor Hill Books, 1978), 85

"Richmond (Echo) Park." https://www.nycgovparks.org/parks/richman-echo-park.

Indian Cave 56

Type of Formation: Rock-Shelter
WOW Factor: Unknown
Location: Spuyten Duyvil (The Bronx)
Tenth Edition, NYS Atlas & Gazetteer: p. 111, C6
Earlier Edition, NYS Atlas & Gazetteer: p. 25, AB4–5
Parking GPS Coordinates: 40°53.144′N, 73°54.896′W
Seton Park GPS Coordinates: 40°53.159′N, 73°54.977′W
Raoul Wallenberg Forest GPS Coordinates: 40°53.248′N, 73°55.071′W
Indian Cave GPS Coordinates: Not determined
Accessibility: >0.1–0.3-mile hike/possibly bushwhack
Degree of Difficulty: Moderate
Additional Information: Seton Park, West 235th Street & Independence Avenue, New York, NY 10463

Description

In *History in Asphalt: The Origin of Bronx Street and Place Names*, John McNamara writes, "Several gorges lead from the steep hillside of the former Seton Hospital grounds atop Spuyten Duyvil. In one of the gorges, some overhanging rocks form a natural cave known locally as 'Indian Cave.'"

History

According to Native American legend, the shelter was occupied by two of Nimham's bands of Stockbridge warriors, taking refuge there after being defeated in a battle near Woodlawn Heights in 1778.

The Raoul Wallenberg Forest is named after Raoul Gustaf Wallenberg, a Swedish diplomat who is credited with saving the lives of thousands of Hungarian Jews during World War II.

Directions

We have done our best to track down the location of this cave site. Although the Seton Hospital was demolished in 1955, we were able to determine that its GPS coordinates were 40°53.100′N, 73°54.933′W, which places the hospital's site 0.1 mile above Seton Park (a small park at an elevation of 184 feet, overlooking the Hudson River, named after Saint Elizabeth Ann Seton).

Downhill from Seton Park is the Raoul Wallenberg Forest, which is our best guess as to where the gorge and shelter cave are likely to be. It should be an easy bushwhack over a very small, defined area to see what can be seen.

If you turn up nothing in the Wallenberg Forest, keep in mind that the forest continues on the other side of Palisade Avenue, all the way down to the Hudson River, now as Riverdale Park. This part of the park might also be worth exploring. A 1.0-mile-long path parallels Palisades Avenue and the Hudson River through this section.

Traveling on the Henry Hudson Parkway, take Exit 19. When you come to West 232nd Street at the end of the ramp, head west for one block. Then turn right onto Independence Avenue and park immediately on the left side of road, which faces Seton Park.

From Seton Park, head west, downhill, towards the Hudson River. Some exploring on your own will be required to find the location of this rock-shelter.

Resources

John McNamara, *History in Asphalt: The Origin of Bronx Street and Place Names* (Harrison, NY: Harbor Hill Books, 1978), 380.

"Raoul Wallenberg Forest." https://www.nycgovparks.org/parks/raoul-wallenberg-forest/history.

Pudding Rock (Historic) 57

Type of Formation: Large Rock
WOW Factor: 7
Location: Bronx
Tenth Edition, NYS Atlas & Gazetteer: p. 111, D6
Earlier Edition, NYS Atlas & Gazetteer: p. 25, B4–5
Pudding Rock Former GPS Coordinates: 40°49.628′N, 73°54.303′W

24. Pudding Rock. *Source*: Harry T. Cook, *The Borough of the Bronx, 1639–1913: Its Marvelous Development and Historical Surroundings* (New York: printed by the author, 1913), 3.

Description

Pudding Rock, aka Puddling Rock, was a prominent landmark until nineteenth-century officials determined that it stood in the way of the city's expansion and destroyed the mammoth rock in the early 1900s.

Stephen Paul Devillo, in an article called "Puddling Rock," describes the rock as "an immense loaf-shaped boulder of sandstone and gravel conglomerate."

History

According to Randall Comfort, in *History of Bronx Borough: City of New York*, "Many are the tales recounted about this huge mass of rock. Rising 'not unlike a puddling in a bag,' it was gracefully ornamented at the top by an attractive group of cedar trees, its dimensions being twenty-five feet high and thirty-five feet in diameter—truly a gigantic boulder in every sense of the word. The Indians of old were not slow in discovering that on one side possessed a natural fire-place, where they cooked their oysters and clams and held their 'corn feasts.'"

The rock was named Pudding Rock because the big, purplish rock was peppered with gravel and small stones, making it look like Christmas plum pudding.

The rock was a frequent resting spot for Protestant French Huguenots as they made their way along Old Boston Road to the French church on John Street in lower Manhattan.

Puddling, the rock's alternate spelling, conjures up a word not commonly used and refers to the process of converting pig iron into wrought iron, subjecting it to intense heat in a furnace. We can't help but wonder if Puddling Rock was possibly a misspelling of Pudding Rock, and yet the name also seems to strangely fit the description of the rock when it was used as a shelter for fires.

According to John McNamara, in *History in Asphalt: The Origin of Bronx Street Names and Place Names*, it's possible that the boulder was also known as Tramp's Rock, serving as a refuge for vagabonds.

It's reported that the decomposed body of bank robber George Leonidas Leslie, aka "Western George" and the "King of Bank Robbers" was found, his head riddled with bullets, in a bush next to Tramp's Rock in 1878.

Directions

Until Pudding Rock was destroyed, it could be seen at the intersection of the Boston Post Road and 166th Street, about a mile west of the Bronx River.

Today, all that you get to see are tall buildings and relentless traffic. Still, if you use your imagination and mentally travel back two centuries, you can almost envision what the area looked like when Pudding Rock, not the tall buildings, was the highest object in the area.

Resources

Randall Comfort, compiler, *History of Bronx Borough: City of New York* (New York: North Side News Press, 1906), 1.

Harry T. Cook, *The Borough of the Bronx, 1639–1913* (New York: printed by the author, 1913), 2. A photograph of the rock can be seen on page 3.

John McNamara, *History in Asphalt: The Origin of Bronx Street Names and Place Names* (Harrison, NY: Harbor Hill Books, 1978), 447.

"The Pudding Rock." https://bronxriver.org/post/story/thepuddingrock.

Pelham Bay Park

Pelham Bay Park, now consisting of 2,772 acres of land, was created in 1888 when New York City bought Hunter Island and Twin Island and some nearby pieces of land and began to combine them into one landmass. It is the largest park in the county and is over three times as large as Central Park.

Naval Camp Boulder 58

Type of Formation: Large Rock
WOW Factor: 5
Location: South of Pelham Manor (Bronx County)
Tenth Edition, NYS Atlas & Gazetteer: p. 111, C7
Earlier Edition, NYS Atlas & Gazetteer: p. 25, AB5
Parking GPS Coordinates: 40°52.174′N, 73°47.749′W
City Island Circle GPS Coordinates: 40°51.598′N, 73°48.114′W
Naval Camp Boulder GPS Coordinates: 40°51.678′N, 73°47.935′W (guesstimate)
Accessibility: 0.3-mile hike/bushwhack from the City Island Traffic Circle
Degree of Difficulty: Moderate
Pelham Bay Map: pelhambaypark.org/trails

Description

The Naval Camp Boulder is a fairly broad rock some 6–7 feet high.

History

The boulder is one of the surface features on the grounds of the former Pelham Bay Naval Camp, which operated from 1917 to 1919. The naval camp was essentially dismantled after World War I.

The rock's existence was recently brought to the forefront again thanks to local history buff Charlie Krieg, who, coming across an old photograph of the rock, induced Jorge Santiago to join him in a search for the boulder. They succeeded, coming across the boulder just northeast of the traffic circle.

Directions

From the Hutchinson River Parkway, take Exit 3 for Orchard Beach/City Island. Drive southeast for 1.0 mile, passing through the Bartow-Pell Traffic

Circle midway. When you come to Park Drive/Orchard Beach Road, turn left and proceed northeast on Park Drive/Orchard Beach Road for 0.4 mile. Turn right into the enormous parking area for Orchard Beach. The parking lot encompasses 4.5 acres of pavement and can accommodate up to 6,888 cars.

From the Orchard Beach pavilion, follow a white-marked trail west for 0.3 mile to reach the bike path that parallels Park Drive/Orchard Beach Road. Walk south for 0.3 mile to reach the City Island traffic circle (whose GPS coordinates have been listed). Follow a white-marked trail northeast into a wooded area bounded by Park Drive/Orchard Beach Road to the west, Orchard Beach and the parking area to the north, Long Island Sound to the east, and City Island Road to the south. This is an area on the Pelham Bay map listed as The Meadow. There are trails in the woods that you can follow, but it may be necessary to bushwhack to locate the boulder. It makes sense to visit when the trees are devoid of leaves, thereby increasing your range of vision dramatically. The GPS coordinates given for the boulder are a guesstimate, but we bet we're not too far off the mark.

Resources

"Boulder Sites: Rocks That Rock." https://www.pelhambaypark.org/boulder-sites. This website shows a photograph, taken about 1918, of a group of naval men standing in front of the boulder.

"Boulder Sites – Friends of Pelham Bay Park." https://www.pelhambaypark.org/boulder-sites. The website has a photograph of the boulder, now grown with trees.

Jack's Rock and Mishow Rock 59

Orchard Beach

Type of Formation: Large Boulder; Historic Rock
WOW Factor: 2
Location: South of Pelham Manor (The Bronx)
Tenth Edition, NYS Atlas & Gazetteer: p. 111, C7
Earlier Edition, NYS Atlas & Gazetteer: p. 25, AB5
Parking GPS Coordinates: 40°52.174′N, 73°47.749′W
Destination GPS Coordinates: *Jack's Rock:* 40°51.748′N, 73°47.614′W; *Mishow Rock*: 40°52.230′N, 73°47.309′W
Accessibility: *Jack's Rock & Mishow Rock*: Distances depend upon where you enter the beach; up to 0.5-mile walk to each
Degree of Difficulty: Easy
Additional Information: *Pelham Bay Map*: pelhambaypark.org/trails

---

Description

Jack's Rock is a large, glacial boulder that lies partially buried on the sandy beach at Orchard Beach. Only its top 2 feet are exposed.

Mishow Rock, aka Wedding Rock, is a large, glacial boulder also in the Orchard Beach area that was nearly buried when Hunter Island was joined to the mainland to create Orchard Beach. According to the City of New York Parks and Recreation's *Pelham Bay Park History*, "only twin points [of the boulder] project from the ground." Reputedly, the rock measured 8 feet high and 12 feet long before its full size was obscured.

In Stephen Jenkins's *Story of the Bronx*, an old photograph shows the rock and a couple of men standing next to it. The rock, with the words "Loreley Point, 1902" painted on it, appears to be 6–7 feet high.

25. Loreley Point Rock. *Source*: Stephen Jenkins, *The Story of the Bronx: From the Purchase Made by the Dutch from the Indians in 1639 to the Present Day* (New York: G. P. Putnam's Sons, 1912), 316–317 insert.

History

Prior to the creation of the 1.0-mile-long, 300–400-foot-wide, crescent-shaped Orchard Beach, Jack's Rock could be seen offshore, a large portion of its bulk rising above the water line.

According to Randall Comfort, in the *History of Bronx Borough: City of New York*, "Jack's Rock [the name of a resort next to the rock was] one of the best fishing resorts in the area." When the sands of Orchard Beach were extended farther out into the ocean, Jack's Rock virtually disappeared except for what shows today.

If Jack's Rock served as the centerpiece for a fishing resort, Mishow Rock, anthropologists believe, earlier served as both a meeting place and ceremonial site for Native Americans.

When Orchard Beach was under construction, which involved bringing in 850,000 cubic yards of brown sand topped by 350,000 cubic yards of fine white sand imported from New Jersey and Rockaway, Mishow Rock would have been completely buried were it not for the efforts of Bronx historian Theodore Kazimiroff, who persuaded developers to leave a portion of the rock exposed for historical preservation.

Directions

Follow the directions given in the Naval Camp Boulder chapter to reach the parking lot for Orchard Beach.

JACK'S ROCK

From the parking lot, walk east to Orchard Beach, turn right, and head south. Look for the boulder near the southeast end of the beach. Keep your eyes open, for the rock will not be a prominent feature of the landscape.

MISHOW ROCK

From the parking lot, walk east to Orchard Beach, turn left, and head toward the northeast end of the beach. When you come to the Kazimiroff Nature Trail, turn left and begin looking for the rock which, from what we've read, lies at the northwest end of the promenade in deep grass.

Resources

New York-New Jersey Trail Conference, *Day Walker: 32 Hikes in the New York Metropolitan Area*. 2nd ed. (Mahwah, NJ: New York-New Jersey Trail Conference, 2002), 54.

"Historic Pelham: The 'Grey Mare' and 'Mishow' Boulders – Part of Pelham's Native American Past." https://historicpelham.blogspot.com/2005/05/grey-mare-and-mishow-boulders-part-of.html.

Randall Comfort, compiler, *History of Bronx Borough: City of New York* (New York: North Side News Press, 1906), 2.

City of New York Parks and Recreation, *Pelham Bay Park History* (New York: Administrator's Office, City of New York Parks and Recreation, 1986), 2.

Bill Twomey, *East Bronx: East of the Bronx River*, Images of America Series (Charleston, SC: Arcadia Publishing, 1999). On page 128 is a photograph of Jack's Rock with Bill Twomey standing behind it. The caption mentions that Skipps Lane goes by the rock.

Bill Twomey, "Do You Remember?," *New York Post*, July 5, 2010, https://nypost.com/2010/07/05/do-you-remember. An article by Bill Twomey provides further details about Jack Rock's history.

Stephen Jenkins, *The Story of the Bronx: From the Purchase Made by the Dutch from the Indians in 1639 to the Present Day* (New York: G. P. Putnam's Sons, 1912). An old photograph of the rock, with two men standing next to it, is shown on pages 316–317.

Raymond H. Torrey, Frank Place Jr., and Robert L. Dickinson, *New York Walk Book*, 3rd ed. (New York: American Geographical Society, 1951), 11. The writers claim that no one knows for sure which rock is Mishow Rock.

Sphinx Rock 60

Twin Island

Type of Formation: Medium-Sized Rock
WOW Factor: 3
Location: Twin Island (The Bronx)
Tenth Edition, NYS Atlas & Gazetteer: p. 111, C7
Earlier Edition, NYS Atlas & Gazetteer: p. 25, B5
Parking GPS Coordinates: 40°52.174′N, 73°47.749′W
Destination GPS Coordinates: *Sphinx Rock*: Not determined;
Several Large Rocks: 40°52.422′N, 73°47.005′W
Accessibility: <0.4-mile hike from northeast end of Orchard Beach
Degree of Difficulty: Moderate
Additional Information: *Pelham Bay Map*: pelhambaypark.org/trails

---

Description

In *The Other Islands of New York City*, Sharon Seitz and Stuart Miller write, "There are several massive boulders, including a glacial erratic known as Sphinx Rock, just before the informal trail bends [from Twin Island] west toward Hunter Island."

Catherine Scott, in "Twin Island: A Bronx Secret," which appeared in the spring 1998 issue of *The Bronx County Historical Society Journal*, writes, "The rock remained here for over 10,000 years with a precariously balanced boulder on top. Within the last several years, however, the top half broke away and now lies near the larger piece. Early colonists noted that the Indians revered this boulder as they did other erratics." It was probably the smaller boulder balanced on top of the larger rock, perhaps looking like a head, that gave rise to the name "Sphinx Rock."

Recent photographs of Sphinx Rock show a highly irregularly shaped rock with both ends above the ground.

Sphinx Rock is perched on the northeastern edge of Twin Island. It is just one of a number of medium-sized boulders that populate the area.

History

In the past, Sphinx Rock was also called Lion Rock, probably because the boulder resembled the Great Sphinx of Egypt, whose body was that of a recumbent lion.

Twin Island (which, until recently, consisted of cigar-shaped East Twin Island and cigar-shaped West Twin Island) became linked with Hunter Island in 1947, just as Hunter Island was connected to Rodman's Neck when Orchard Beach was created in the 1930s.

Directions

Follow the directions given in the Naval Camp Boulder chapter to reach the parking lot for Orchard Beach.

From the northeast end of the parking area, walk east and then northeast along the Beach Promenade for >0.3 mile to reach the northeast end of Orchard Beach. From the nature center, follow a white-marked trail northeast that leads to the tip of Twin Island, a hike of 0.4 mile. Twin Island, once an island, today is simply a grassy extension of the northeast tip of Orchard Beach. Look for large erratics along the way. One of them will be the historic Sphinx Rock.

Resources

Sharon Seitz and Stuart Miller, *The Other Islands of New York City: A History and Guide* (Woodstock, VT: Countryman Press, 2001), 125.

Catherine Scott, "Twin Island: A Bronx Secret," *Bronx County Historical Society Journal* 35, no. 1 (Spring 1998): 31.

Leslie Day, *Field Guide to the Natural World of New York City* (Baltimore, MD: Johns Hopkins University Press, 2007), 27– 28.

"Hunter and Twin Islands – Friends of Pelham Bay Park." https://www.pelhambaypark.org/hunter-and-twin-islands. Photograph of Sphinx Rock by John Grayley.

Tillie's Rock 61

Hunter Island

Type of Formation: Large Boulder
WOW Factor: 4
Location: Hunter Island (Bronx County)
Tenth Edition, NYS Atlas & Gazetteer: p. 117, C7
Earlier Edition, NYS Atlas & Gazetteer: p. 25, B5
Parking GPS Coordinates: 40°52.174'N, 73°47.749'W
Tillie's Rock GPS Coordinates: 40°52.645'N, 73°47.062'W and 40°52.614'N, 73°47.025'W
Accessibility: 0.5-mile hike from north end of Orchard Beach
Degree of Difficulty: Moderate
Additional Information: *Pelham Bay Map*: pelhambaypark.org/trails

---

Description

Tillie's Rock, aka Tilly's Rock, is the last rock in a series of medium-sized boulders along a narrow strip of land extending out from the northeast tip of Hunter Island.

History

Years ago, the Tillie's Rock area grew to be a favorite bathing site for men and boys since, due to its inaccessibility, it provided seclusion for nude bathing (which is why, obviously, women chose not to go there). This privacy ended when Orchard Beach was created and Hunter Island was joined to the mainland.

No one knows for sure how the rock was named. One theory is that the rock was originally named Tiller Rock by seamen who used it as a navigational marker to take tiller and change their course. A second theory is that the name arose from a woman named Matilda.

Siwanoy Native Americans called Hunter Island "Laap-Ha-Wach-King," meaning "the place of stringing beads." While most have taken this to mean the stringing together of shells for wampum, we can't help but wonder if it actually refers to the bead-like string of boulders at Tillie's Rock.

Directions

Follow the directions given in Naval Camp Boulder chapter to reach the parking lot for Orchard Beach.

From the northeast end of the Orchard Beach Promenade, follow the red-marked Kazimiroff Trail and spur paths northeast until you come to the northeast end of the island where sizeable boulders can be seen both along the ridge line and along a 0.1-mile-long spit of land attached to the island. During high tide, this 120-foot-long thread of land disappears underwater, leaving Tillie's Rock isolated on an island.

Undoubtedly, the bathing area that attracted so many swimmers prior to modern times was in this general section of Tillie's Rock.

Resources

John McNamara, *History in Asphalt: The Origin of Bronx Street and Place Names* (Harrison, NY: Harbor Hill Books, 1978), 126–127, 487.

"Tillie's Rock, A Swimming Hole Paradise for the Boys and Men of Pelham." https://historicpelham.blogspot.com/2015/07/tillies-rock-swimming-hole-paradise-for.html.

Grey Mare Rock 62

Hunter Island

Type of Formation: Large Rock
WOW Factor: 4
Location: Hunter Island (The Bronx)
Tenth Edition, NYS Atlas & Gazetteer: p. 117, C7
Earlier Edition, NYS Atlas & Gazetteer: p. 25, AB5
Parking GPS Coordinates: 40°52.174′N, 73°47.749′W
Gray Mare Rock GPS Coordinates: 40°52.857′N, 73°47.371′W (Google Earth)
Accessibility: 1.0-mile hike
Degree of Difficulty: Moderate
Additional Information: *Kazimiroff Nature* **Trail Map:** nycgovparks.org/sub_about/parks_divisions/nrg/documents/NRG_Publication_The_Kazimiroff_Nature_Trail_Pelham_Bay_Park_Bronx.pdf

---

Description

Grey Mare Rock is a large, grayish, moss-covered boulder that was brought to its present position by glaciers during the last Ice Age. It has also been spelled "Gray Mare" at various times.

In *Field Guide to the Natural World of New York City*, Leslie Day writes, "You will see a large boulder, or glacial erratic protruding from the water. The Siwanoy Indians called this sacred ceremonial site the Grey Mare."

Interestingly, the term Grey Mare (as applied to a rock) originated in Northumberland, England, where locals used the word to describe boulders that tapered at the top. Such rocks attracted children, who would climb up and straddle them, and then pretend to be riding a horse.

History

In 1881, Robert Bolton Jr., wrote of the Grey Mare, "To this piece of rude natural sculpture, the Indians invariably paid just respect, believing it to have been placed there by the direct interposition of their God or guardian Manito, for their especial benefit or favor."

According to Sharon Seitz and Stuart Miller in *The Other Islands of New York City*, "Hunter was called *Laap-Ha-Wach-King*, or 'place of Stringing Beads' by the Siwanoy, who held religious rituals on the island's odd rock formations." The island has gone by other names over time, including Pells Island (1654), Pelican Island, Appleby's Island (during the Revolutionary War), Blagge's Island, and Henderson Island, a number of these reflecting the names of previous landowners.

The name Hunter Island stuck after John Hunter purchased the property in 1804 and built his grand mansion on it in 1813. The house survived through multiple owners until it was destroyed in 1937 at the time when Orchard Beach was created.

The Kazimiroff Nature Trail, which opened in 1986, was named for past noted Bronx naturalist Dr. Theodore Kazimiroff. It takes you to various sections of the (originally) 215-acre Hunter Island.

Directions

Follow the directions given in the Naval Camp Boulder chapter to reach the parking lot for Orchard Beach.

Both the red-marked and blue-marked Kazimiroff Trails can be accessed from the northeast end of the Orchard Beach Promenade. The blue-marked trail will get you closest to the northwest end of the island. From there, you will need to continue walking toward the island's northern tip via secondary trails, where Grey Mare Rock can be found on marshy lands facing the Long Island Sound.

Resources

New York-New Jersey Trail Conference, *New York Walk Book*, 6th ed. (New York: New York-New Jersey Trail Conference, 1998), 53.

"The Kazimiroff Nature Trail." https://www.nycgovparks.org/sub_about/parks_divisions/nrg/documents/NRG_Publication_The_Kazimiroff_Nature_Trail_Pelham_Bay_Park_Bronx.pdf

New York-New Jersey Trail Conference, *Day Walker: 32 Hikes in the New York Metropolitan Area*, 2nd ed. (Mahwah, NJ: New York-New Jersey Trail Conference, 2002), 50–52.

"Historic Pelham: The 'Grey Mare' and 'Mishow' Boulders – Part of Pelham's Native American Past." https://historicpelham.blogspot.com/2005/05/grey-mare-and-mishow-boulders-part-of.html.

"The Named Stones of Northumberland (revisited)." http://heddonhistory.weebly.com/blog/the-named-stones-of-northumberland-revisited.

Sharon Seitz and Stuart Miller, *The Other Islands of New York City: A History and Guide*, 3rd ed. (Woodstock, VT: Countryman Press, 2011), 120, 130–135.

Leslie Day, *Field Guide to the Natural World of New York City* (Baltimore, MD: Johns Hopkins University Press, 2007), 31.

"Boulder Sites – Friends of Pelham Bay." https://www.pelhambaypark.org/boulder-sites.63

Glover's Rock

Pelham Bay Park

Type of Formation: Large Boulder
WOW Factor: 3–4
Location: Pelham Manor (The Bronx)
Tenth Edition, NYS Atlas & Gazetteer: p. 111, C7
Earlier Edition, NYS Atlas & Gazetteer: p. 25, AB5
Parking GPS Coordinates: *Split Rock Golf Course*: 40°52.319′N, 73°48.617′W; *Orchard Beach*: 40°52.103′N, 73°47.860′W
Destination GPS Coordinates: *Glover's Rock*: 40°51.902′N, 73°48.202′W; *Boulder near Glover's Rock*: 40°51.895′N, 73°48.202′W
Accessibility: Roadside or 0.6-mile walk
Degree of Difficulty: Moderately easy
Pelham Bay Map: pelhambaypark.org/trails

Description

Glover's Rock is an 8-foot-high, granite boulder of historical significance.

A second, sizeable erratic lies 30 feet behind Glover's Rock.

History

Glover's Rock commemorates the Battle of Pell's Point when Colonel John Glover in October of 1776 led a small brigade of 750 Americans against British General William Howe's force of 4,000 redcoats and Hessians. Glover's forces didn't win the battle, but strategically delayed the British's advancement just long enough to allow General George Washington and his men to escape to White Plains, where they were able to rest and regroup.

There is an old account that suggests that Colonel Glover stood on top of the rock to track the British forces as they landed. This, undoubtedly, is apocryphal, but remains a lively story.

A plaque was installed on the rock in 1901 by the Daughters of the American Revolution to honor the 125th anniversary of the Battle at Pell's Point and Colonel Glover's victory. Vandals ripped it off, but a new, larger bronze tablet that was installed in 1960 still remains in place. A sign next to the rock provides valuable historical information.

Directions

Glover's Rock is located on Orchard Beach Road, less than 0.5 mile southeast of the Bartow-Pell traffic circle.

Driving on the Hutchinson River Parkway, take Exit 3 for Orchard Beach/City Island. When you come to the Bartow-Pell traffic circle after 0.4 mile, head northeast on Shore Road for 0.3 mile to reach the Split Rock Golf Course, on your left.

Starting from the Golf Course's parking area, pick up the yellow-marked Siwanoy Trail and follow it southeast. The path initially follows along Shore Road and then, after going under Orchard Beach Road, takes you southeast, paralleling Orchard Beach Road. In 0.7 mile, you will come Glover's Rock, on a white-marked spur path to your left.

The rock can also be accessed from the Siwanoy Trail, starting at a wooded area called The Meadows by Orchard Beach.

The easiest way if you don't want to include a hike, however, is to simply turn into a pull-off on your right next to the rock. Note: This only works if you are traveling on the east-bound lane of Orchard Beach Road.

Resources

"Pelham Bay Monuments – Glover's Rock." https:/www.nycgovparks.org/parks/pelham-bay-park/monuments/590.

"Glover's Rock Historical." https://historicalmarkerproject.com/markers/HM1UZ3_glovers-rock-historical_NY.html.

Catherine A. Scott, *City Island and Orchard Beach* (Charleston, SC: Arcadia Publishing, 1999). On page 11 is a photograph of Glover's Rock.

"Bronx Revolutionary Remains." https://forgotten-ny.com/2000/08/bronx-rocks-revolutionary-war-remains-in-pelham-bay-park.

"Glover's Rock." https://www.bridgeandtunnelclub.com/bigmap/bronx/pelhambaypark/gloversrock/index.htm. Several photographs of Glover's Rock are shown on this website.

Robert F. Ryan, "John Glover and the Battle of Pell's Point," *Bronx County Historical Society Journal* 2, no. 2 (July 1965): 65–85. Information on the rock is provided on pages 67 and 68. On page 67 is a full account of the fate of the Glover's Rock plaque.

John McNamara, *McNamara's Old Bronx* (Harrison, NY: Harbor Hill Books, 1978), 241. A photograph of the rock is shown, along with some early members of the Bronx Historical Society.

Stephen Jenkins, *The Story of the Bronx: From the Purchase Made by the Dutch from the Indians in 1639 to the Present Day* (New York: G. P. Putnam's Sons, 1912), 144–145. An insert photo shows Glover's Rock, including a large boulder nearby in the background. On page 311 is the inscription contained on an old plaque on Glover's Rock.

Split Rock and Eagle Rock 64

Pelham Bay Park

Type of Formation: Split Rock; Historic Rock
WOW Factor: 6
Location: Pelham Manor (The Bronx)
Tenth Edition, NYS Atlas & Gazetteer: p. 111, C7
Earlier Edition, NYS Atlas & Gazetteer: p. 25, AB5
Parking GPS Coordinates: 40°52.319′N, 73°48.617′W
Bartow Circle GPS Coordinates: 40°52.133′N, 73°48.642′W
Destination GPS Coordinates: *Split Rock*: 40°53.187′N, 73°48.893′W; *Eagle Rock*: Unknown
Accessibility: *Split Rock*: >1.0-mile hike; *Eagle Rock*: Unknown
Degree of Difficulty: Moderate
Pelham Bay Map: pelhambaypark.org/trails

Description

Split Rock, as the name suggests, is a 12–15-foot-high, 25-foot-long glacial boulder that has cracked into two halves. The gap between the two halves is fairly wide and would be sufficient to conceal a person or persons except that a tree has grown up in it over the years.

Randall Comfort, in *History of Bronx Borough: City of New York*, says of Split Rock, "Cleft directly in the middle with a good-sized tree growing in the fissure, this great boulder is one of the sights of the neighborhood." Or at least it was one of the neighborhood sights during Randall Comfort's time. Today, the rock is partially isolated by superhighways—the Hutchinson River Parkway and New England Thruway—that pass right by it.

Most people driving along the Hutchinson River Parkway simply have no idea that the highway came to be named for Anne Hutchinson, or how her history is intimately associated with Split Rock.

As it is, the historic rock barely made it into modern times. In the 1960s, Split Rock was almost blasted away by engineers while clearing the route for Interstate 95. Fortunately, community activism initiated by the Bronx Historical Society prevented the boulder from being demolished, and the Hutchinson River Parkway was subsequently moved 50 feet to the north.

History

In 1643, Anne Hutchinson [see chapter on Helicker's Cave for more information on Anne Hutchinson's exploits] and members of her family hid in the crevice at Split Rock to avoid being captured by Siwanoy Indians. Apparently, this ruse ultimately failed, for Ann and her family were captured and subsequently massacred—the one possible exception being Ann's daughter, Susanna, whose fate is not known for certain.

A tablet was affixed to the boulder in 1911 by the Colonial Dames of the State of New York to honor Anne Hutchinson. By 1914, however, the tablet was stolen by vandals.

Fortunately, we know that the plaque read, "Anne Hutchinson. Banished from the Massachusetts Bay Colony in 1638 because of her devotion to religious liberty this courageous woman sought freedom from persecution in New Netherland. Near this rock in 1643 she and her household were massacred by Indians. This tablet is placed here by the Colonial Dames of the State of New York. Anno Domini MCMXI [1911] Virtutes Majorum Filiae Conservant."

In *The Shaping of North America*, Isaac Asimov contends that Anne Hutchinson "was the first woman of note in American history."

Recent concerns have been raised about the rock breaking loose from the supporting cliff base and falling onto the highway below. Presumably, action can be (or has been) taken to prevent this from happening.

EAGLE ROCK

Although not a boulder, but rather a bluff, Eagle Rock is mentioned by John McNamara in his book *History in Asphalt: The Origin of Bronx Street Names and Place Names* and is described as "a prominent bluff on east bank of the Hutchinson River 200 yards north of the Hudson River Parkway." We have no idea if the bluff still exists today.

Directions

From the Hutchinson River Parkway, take Exit 3 for Orchard Beach/City Island. When you come to the Bartow-Pell traffic circle after 0.4 mile, head northeast on Shore Road for 0.3 mile to reach the Split Rock Golf Course. Turn left to park. If you enjoy golf, you may want to play a few rounds while visiting.

From the golf course parking lot, walk south along the bike path to the Bartow-Pell traffic circle, and then follow the Split Rock Trail as it meanders through Goose Creek Marsh and then the 50-acre Thomas Pell Sanctuary. The trail was restored in 1985 by the park and the Mayor's City Volunteer Corps.

If you look at a Pelham Bay map, you will notice that Split Rock can also be reached by taking the Bridle Path from the west side of the parking lot that goes north to the northwest corner of Pelham Bay Park, near where Split Rock is located, before circling back.

Split Rock can be seen from both the Split Rock Trail and Bridle Path, which are enjoined at this point. The problem, however, is that Split Rock is on a tiny triangle of land walled off by two super-highways— the Hutchinson River Parkway and the New England Thruway. The only way to physically get to the rock is to cross over a busy exit ramp from the Hutchinson River Parkway. Doing this is not something that is recommended. For this reason, you may want to content yourself with a view of the rock from the east side of the exit ramp.

However, it would seem that the rock can be seen from the superhighways. In Mary Andrews's article on Split Rock, she states, "The rock stands beside the throughway where it passes under the Hutchinson Parkway and may be seen from either highway."

Resources

Catherine A. Scott, *City Island and Orchard Beach* (Charleston, SC: Arcadia Publishing, 1999). On page 10 is a photograph of the boulder with four men standing on top of it.

Stevenson Swanson, "Who Was Anne Hutchinson?" *Patch* (Larchmont-Mamaroneck, NY), June 10, 2011, https://patch.com/new-york/larchmont/who-was-anne-hutchinson-2.

Randall Comfort, compiler, *History of Bronx Borough: City of New York* (New York: North Side News Press, 1906), 1.

City of New York Park and Recreation, *Pelham Bay Park History* (New York: Administrator's Office, City of New York Park and Recreation, 1986). A photograph of Split Rock can be seen on page 4.

Westchester Historical Society, *Anne Hutchinson and Other Papers* (White Plains, NY: Westchester County Historical Society, 1929). The rock is described as "a great cloven boulder." A photograph of the rock atop a hill as seen from a great distance is shown on an insert between pages 12 and 13.

John McNamara, *History in Asphalt: The Origin of Bronx Street and Place Names* (Harrison, NY: Harbor Hill Books, 1978), 216.

Stephen Jenkins, *The Story of the Bronx* (New York: G. P. Putnam's Sons, 1912). On an insert between pages 310 and 311 is a photograph of Split Rock.

Harry Hansen, *North of Manhattan: Persons and Places of Old Westchester* (New York: Hastings House, 1950). Detailed information about Anne Hutchinson is provided on pages 137–165. On page 150, the author writes, "Across the river and on the eastern slope, directly above the new parkway, stands an immense double boulder, the split rock that became the first landmark for supervisors in this area."

Harry T. Cook, *The Borough of the Bronx, 1639–1913* (New York: printed by the author, 1913), 3. On pages 58 and 123 are photographs of Split Rock.

"Bronx Revolutionary Remains." https://forgotten-ny.com/2000/08/bronx-rocks-revolutionary-war-remains-in-pelham-bay-park.

Isaac Asimov, *The Shaping of North America* (London: Dobson Books, 1973), 105.

Raymond H. Torrey, Frank Place Jr., and Robert L. Dickinson, *New York Walk Book*, 3rd ed. (New York: American Geographical Society, 1951), 12. Split Rock and the story of Anne Hathaway are discussed.

Mary Andrews, "Split Rock," *Westchester Historian: Quarterly of the Westchester County Historical Society* 38, no. 2 (April–June 1962): 54. A stark photograph of the rock, taken by Mary Andrews, is shown on page 52.

Roosevelt Rock 65

Pelham Bay

Type of Formation: Medium-Sized Rock; Historic Rock
WOW Factor: 3
Location: Pelham Manor (The Bronx/Westchester County)
Tenth Edition, NYS Atlas & Gazetteer: p. 111, C7
Earlier Edition, NYS Atlas & Gazetteer: p. 25, AB5
Shore Road & Pelhamdale Ave. Junction GPS Coordinates: 40°53.218'N, 73°47.469'W
Parking GPS Coordinates: 40°52.890'N, 73°47.718'W
Roosevelt Rock GPS Coordinates: 40°52.955'N, 73°47.605'W (estimated)
Accessibility: 0.1-mile walk
Degree of Difficulty: Moderately easy

Description

Roosevelt Rock is a medium-sized, fairly nondescript rock with historic significance that faces Roosevelt Cove. It may take some effort to locate the rock, which looks undistinguished, and to then find an inscription chiseled into it.

History

The boulder is located on the former grounds of the Elbert Cornelius Roosevelt estate. Roosevelt was a New York City merchant and distantly related to the famous Roosevelt family. One of Elbert's sons, Isaac, made a carving in the rock in 1833 that read, "Isaac Roosevelt. 1833." Apparently, the words are still legible nearly two centuries later.

Readers may be interested to know that this is not the first Roosevelt Rock that we have encountered or written about. In *Boulders Beyond*

Belief: An Explorer's Hiking Guide to Amazing Boulders and Natural Rock Formations of the Adirondacks (Chapter 97), we describe an historic rock associated with John Ellis Roosevelt, cousin of Theodore Roosevelt.

Directions

The Historic Pelham website, historicpelham.blogspot.com/2006/11, dated November 13, 2006, provides fairly clear directions for accessing Roosevelt Rock:

> As you leave Pelham on Shore Road [heading south] you will pass Shore Park on the left and a number of homes on the east side of Shore Road. Shortly after you leave Pelham and enter Pelham Bay Park, there is a small parking area on the left (east side) of the roadway. Its entrance usually is blocked with boulders. Several footpaths are accessible from that parking area. They lead down to the shoreline.
>
> At low tide it is easy to walk along the shoreline back toward the Pelham Town Boundary (northward). You will reach the "end" of the shoreline where a fence blocks your continued progress along portions of the shore owned by private home owners. The last large boulder lying on the shore at that spot contains the carving, although it is difficult to see.

Our interpretation of these directions, starting from the junction of Shore Road and Pelhamdale Avenue (which is 0.2 mile north of Shore Park), is to drive southwest on Shore Road for <0.5 mile. Then turn left and park in a small area where boulders block a more substantial parking area off in the woods.

We would suggest, however, approaching from Pelham Bay Park. Get off from the Hutchinson River Parkway at Exit 3. At the Bartow-Pell traffic circle, head north on Shore Road for 1.3 miles, and park to your right in a small space between two guardrails.

Follow a path southeast for 200 feet to reach the shoreline. From here, walk north along the beach for < 0.1 mile until you come to the rock.

Resources

"Elbert Roosevelt, an Early Settler of the Manor of Pelham, and Other Members of His Family," *Historic Pelham*. https://historicpelham.blogspot.com/2014/05/elbert-roosevelt-early-settler-of-manor.html.

"The Isaac Roosevelt Stone Carved in 1833," *Historic Pelham*. https://historicpelham.blogspot.com/2006/11/isaac-roosevelt-stone-carved-in-1833.html.

"Boulder Sites: Rocks That Rock." https://pelhambaypark.org/boulder-sites. This website shows a photograph of Roosevelt Rock.

John McNamara, *History in Asphalt: The Origin of Bronx Street and Place Names* (Harrison, NY: Harbor Hill Books, 1978), 455.

Indian Prayer Rock

66

Athletic Field

Type of Formation: Large Rock
WOW Factor: 8
Location: Pelham Manor (The Bronx)
Tenth Edition, NYS Atlas & Gazetteer: p. 117, C7
Earlier Edition, NYS Atlas & Gazetteer: p. 25, B5
Parking GPS Coordinates: 40°50.922′N, 73°49.287′W
Indian Prayer Rock GPS Coordinates: 40°51.123′N, 73°49.197′W (estimated)
Accessibility: 0.2–0.3-mile walk
Degree of Difficulty: Easy

26. Indian Prayer Rock. Antique postcard, public domain.

Description

Indian Prayer Rock, aka Indian Rock, is a large, black-colored rock formation that may have initially been one huge, intact boulder before geological forces shattered it into smaller, but still large, pieces. The rock formation, made out of gneiss and schist, rises up 25 feet at its highest point. Look closely and you will see igneous intrusions of quartz and feldspar in the rock.

In 1913, a bronze plaque was affixed to the rock to celebrate the 30th anniversary of the creation of the Bronx parks system, but vandals pried it off and stole it many decades ago.

History

Indian Prayer Rock was named for its past association with the Siwanoy and the Lenape who frequented the boulder for religious ceremonies. Unlike the Incas and Mayas, these northeast Native Americans never used rocks to construct massive stone structures; rather, they simply used what was at hand and were satisfied to leave it the way it was found.

Indian Prayer Rock has been utilized in different ways during the twentieth century. In 1904, in celebration of the opening of the new athletic field and parade grounds, the rock was used as a backdrop for archery practice, thus ensuring that errant arrows, missing their mark, would hit the boulder and be harmlessly deflected.

Later, a grandstand was placed next to Indian Prayer Rock, which provided an enclosure or side wall for one of the bleachers. The rock was so close, in fact, that it is said that spectators could reach out and touch it.

Directions

Driving north on the I-95/Buckner Expressway, take Exit 7C for Pelham Bay/Country Club Road. Cross over County Club Road and continue north on Bruckner Boulevard/Macdonough Avenue for >0.2 mile. Turn right onto Middletown Road and drive northeast for 0.3 mile. When you come to Stadium Avenue, turn left and proceed to the parking area.

Driving south on the I-95/Bruckner Expressway, take Exit 8A for Westchester Avenue. Continue south on Bruckner Boulevard for 0.8 mile. When you come to Country Club Road, turn left and drive southeast for 0.1 mile. Then turn left onto Macdonough Avenue and head north for >0.2 mile. Finally, turn right onto Middletown Road and drive northeast for 0.3 mile. When you come to Stadium Avenue, turn left and proceed to the parking area.

According to directions from Blake A. Bell, Pelham historian, Indian Rock is "located in a wooded area just behind the baseball diamond after you pass the dog run. To find the site, walk into the woods from center field." Another source says that the rock's "site is located in the southwestern section of Pelham Bay Park beyond Pelham Bridge and the reclaimed and rehabilitated Bronx-Pelham Landfill. It is in a wooded area that can be entered from center field of the baseball diamond near the dog run." We suspect these directions will make sense once you are at the park.

Our best guess is to walk north from the Aileen B. Ryan Recreation Complex parking area for >0.1 mile, passing by the Aileen B. Ryan Recreational Center to your left. Turn right onto Dog Run (Road) and head northeast for >0.1 mile. You will see a baseball diamond to your left.

Walk out to center field and begin looking for the rock near or back from the edge of the forest.

Resources

"Indian Prayer Rock in Pelham Bay Park." https://historicpelham.blogspot.com/2017/05/indian-prayer-rock-in-pelhampark.html.

"Indian Prayer Rock. The Bronx Chronicle."
https://thebronxchronicle.com/2014/09/30/indian-prayer-rock.

"I Love the Bronx – Boulders Amos el Bronx." https://thebronxfreepress.com/i-love-the-bronx-boulders-amo-el-bronx-rocas. A photograph of the rock is shown on the website.

Nassau County

Nassau County is situated on western Long Island and is the most densely populated county in the state. Its name comes from the Dutch family of King William III of England, the House of Nassau.

The county encompasses 453 square miles, of which 285 square miles is land.

Execution Rock 67

Type of Formation: Historic Rock
WOW Factor: 2–3
Location: Larchmont (Nassau County/Westchester County)
Tenth Edition, NYS Atlas & Gazetteer: p. 111, C8
Earlier Edition, NYS Atlas & Gazetteer: p. 25, AB6
Execution Rock GPS Coordinates: 40°52.681′N, 73°44.261′W
Evers Marina GPS Coordinates: 40°50.688′N, 73°48.895′W
Accessibility: >6.0-mile boat trip from Eastchester Bay
Degree of Difficulty: Easy by boat
Additional Information: Evers Marina, 1470 Outlook Avenue, Bronx 10465; (718) 863-9111

27. Execution Rock Lighthouse. Antique postcard, public domain.

Description

A field of small-to-medium-sized rocks and boulders virtually surrounds Execution Rock and its lighthouse. The boulders would undoubtedly go unnoticed were it not for the fact that legend and myth also surround the rocks, just as the rocks, in turn, surround the lighthouse.

History

According to legend, the boulders along the island achieved notoriety when British executioners chained condemned prisoners to the rocks during low tide, causing then to drown as the tide came in—a very slow, agonizing death indeed, both physically and psychologically (assuming it is true, of course).

It's entirely more likely, however, that the island and its submerged rocks got its name from posing a serious hazard to passing ships, doing more than its share to drown those unfortunate enough to run into the low-lying rocks. Folks at nearby Sands Point (then known as Cow Neck) called the island Executioner's Rock.

The granite, 55-foot-high lighthouse on Execution Rock was constructed in 1849, followed by the lightkeeper's house in 1867. The lighthouse was automated in 1979, no longer requiring personnel living on the island to maintain it.

We have read that Execution Rock is now a bed-and-breakfast for tourists looking for a once-in-a-lifetime experience. My wife, Barbara Delaney, and I stayed overnight at the Saugerties Lighthouse on the Hudson River years ago, and this kind of experience is one that no one should miss out on.

Directions

Execution Rock lies 1.0 mile across the water from the Sands Point shoreline. It is visible from the shore, but obviously at some distance.

Realistically, the only way to access Execution Rock is by water. One way of doing so is to take a 6.0-mile boat trek from Evers Marina, which is located in Eastchester Bay.

From Eastchester (junction of Griswold Road and Macdonough Place) in Pelham Bay, drive east on Griswold Avenue for 0.5 mile, and then turn right onto Outlook Avenue. Evers Marina will be immediately on your left.

Note: This is in the general area of Prayer Rock (see chapter on Prayer Rock).

ADDITIONAL OFFSHORE ROCKS

There are many named rocks in the waters around New York City, Long Island, and Staten Island. The following represent a mere sample of possibilities if you wish to head out and visit them by boat: Jones Rocks (40°59.301′N, 73°38.092′W), Grassy Rocks (40°59.438′N, 73°38.851′W), Channel Rocks (40°59.125′N, 73°39.235′W), Great Captain Rocks (40°59.030′N, 73°39.025′W), Red Rocks (41°00.391′N, 73°36.746′W), Pecks Rocks (41°00.818′N, 73°36.201′W), Hitchcock Rocks (41°00.632′N, 73°36.259′W), Salt Rock (41°00.617′N, 73°35.817′W), Cove Rock (41°00.528′N, 73°35.331′W), Woolsey Rock (41°00.031′N, 73°33.912′W), Highwater Rock (41°01.116′N, 73°32.518′W), Flathead Rock (41°01.396′N, 73°32.772′W), Bold Rock (41°01.833′N, 73°29.584′W), Yellow Rock (41°02.960′N, 73°24.872′W), Old Baldy (41°02.698′N, 73°25.544′W), Copps Rocks (41°03.558′N, 73°22.855′W), Beers Rocks (41°03.857′N, 73°22.760′W), Chimon Rock (41°04.015′N, 73°22.921′W), Dunder Rock (41°04.626′N, 73°20.942′W), Haycock Rock (41°04.834′N, 73°21.498′W), Big Boil (41°14.115′N, 72°54.148′W), and The Chimney (41°14.186′N, 72°53.755′W).

Resources

"A Trip to Execution Rocks: New York's Most Unusual B&B." https://www.scoutingny.com/a-trip-to-execution-rocks.

"The Creepy, Bloody History of Execution Rocks." https://knowledgenuts.com/execution-rocks/.

Richard M. Lederer Jr., *The Place-Names of Westchester County, New York* (Harrison, NY: Harbor Hill Books, 1978), 47.

Frances Meyer Lawrence, "The Story of a Rock," *Nassau County Historical Journal* 15, no. 1 (Spring 1954): 21–25.

Sands Point Preserve Boulders 68

Sands Point Preserve

Type of Formation: Medium-Sized Boulder
WOW Factor: 5
Location: Manorhaven (Nassau County)
Tenth Edition, NYS Atlas & Gazetteer: p. 111, C8
Earlier Edition, NYS Atlas & Gazetteer: p. 25, B6
Parking at Castle Gould GPS Coordinates: 40°51.693′N, 73°41.908′W
Destinations GPS Coordinates: *Sands Point Giant*: Not determined; *Boulders in general*: 40°51.822′N, 73°41.824′W and at 40°51.622′N, 73°41.147′W; *Pond Boulder*: 40°51.740′N, 73°41.810′W; *Hen Boulder with 6 Chicks*: 40°51.623′N, 73°41.048′W (guesstimate)
Fee: Modest admission charged
Accessibility: 1.0-mile trek along beach
Degree of Difficulty: Moderately easy
Additional Information: Sands Point Preserve Conservancy, 127 Middle Neck Road, Sands Point, NY 11050; (516) 571-7901.
Map of Preserve: sandspointpreserveconservancy.org/wp-content/uploads/2017/06/Sands-Point-Preserve-Map.pdf
Guided meditative Forest Bathing Walks offered

Description

Dozens of large, 20-foot-high boulders are contained in the 216-acre Sands Point Preserve.

Sands Point Giant is a large, 17-foot-high, 20–40-foot-long boulder that is located on a private estate at Sands Point. It was first reported by Samuel L. Mitchill (a lecturer on botany, zoology, and mineralogy at Columbia College) in 1800.

Hen Boulder (49 feet by 39 feet by 8 feet) with 6 Chicks (12–16 feet) are rocks located just offshore southeast of "Falaise" (the historic Guggenheim mansion), lying in shallow water. Unfortunately, they are only visible at low tide. Geologists believe that the rocks were at one time one large supergiant until the boulder broke up.

History

Sands Point is named for Captain John Sands, who visited the area in 1695.

The preserve is the former estate of railroad magnate Jay Gould, who purchased the land in 1900–1901. He subsequently built two enormous mansions on the property in an effort to please his wife, Katherine Clemmons. In this he was not successful, for the two divorced after Clemmons allegedly had an affair with William F. Cody, better known as Buffalo Bill.

The land was subsequently purchased in 1917 by mining tycoon Daniel Guggenheim. His son, Harry, built a mansion on a bluff overlooking Long Island Sound and called his estate "Falaise" (French for "cliff").

When Daniel Guggenheim died in 1930, his wife, Florence, built a smaller mansion on the property and called it "Mille Fleurs," French for "a thousand flowers."

It seems fair to say that this was truly a family of mansion builders.

The story of ownership and usage gets somewhat complicated from here (as if it wasn't already), so we'll just skip to the part where Nassau County acquired 127 acres of prime land in 1971, including the former Gould/Guggenheim estate, which became a museum. All of this, today, is the Sands Point Preserve.

Directions

Driving west along the Long Island Expressway (I-495), take Exit 36 for Searingtown Road/Rock Shelter Road and at a traffic light, turn right onto Route 101/Searingtown Road.

Driving east along the Long Island Expressway (I-495), take Exit 36 for Searingtown Road/Port Washington, and then turn left onto Router 101/Searingtown Road.

Head north on Route 101/Searingtown Road/Port Washington Boulevard/Middle Neck Road for 6.0 miles. Eventually, you will see signs for the Sands Point Preserve. Turn right and park in the area for Castle Gould.

To reach the Pond Boulder, walk east from the parking lot. You will quickly come to the pond, where a good-sized boulder can be seen along the east side.

From here, a road leads northwest down to the beach, where you can walk southeast along the shoreline for 1.0 mile, passing by a number of medium-to-large-sized boulders. We've taken a guess and assigned a GPS to where we think the Hen Boulder with 6 Chicks is located.

The Sands Point Giant, unfortunately, is on private land and inaccessible.

Resources

"Sand Point Preserve." http://sandspointpreserveconservancy.org.
"Mission and History – Sands Point Preserve." http://sandspointpreserve conservancy.org/about/mission-history.

Council Rock 69

Type of Formation: Large Boulder
WOW Factor: 3
Location: Oyster Bay (Nassau County)
Tenth Edition, NYS Atlas & Gazetteer: p. 111, C10
Earlier Edition, NYS Atlas & Gazetteer: p. 25, AB7
Council Rock GPS Coordinates: 40°52.427′N, 73°32.480′W
Accessibility: Roadside
Degree of Difficulty: Easy

28. Council Rock. Public domain.

Description

Council Rock is a medium-to-large-sized boulder that lies across the road from 0.2-mile-long Mill Pond, just uphill from Lake Avenue. A historic marker, erected next to the rock in 1939, tells the story of the boulder's history.

History

Council Rock was the location of a sacred council fire and a gathering place for the Matinecock (a tribe of skilled hunters and fishermen).

George Fox, who was the founder of the Society of Friends, is said to have used this huge rock in 1672 to preach at a four-day-long meeting.

An inscription on the rock reads, "Council Rock / Here George Fox 1672 met with / Wrights, Underhill and Feeke / at a Quaker Meeting"—the last names probably referring to Captain John Underhill and his wife, Elizabeth Feake (spelled slightly differently on the plaque).

Directions

At the village of Oyster Bay (junction of West Main Street & South Street), drive northwest on West Main Street for 0.6 mile. When you come to Lake Avenue, turn left and head south for 300 feet. Council Rock is on the right side of the road, just uphill from the historical marker.

Resources

Raymond E. Spinzia, Judith A. Spinzia, and Kathryn E. Spinzia, *Long Island: A Guide to New York's Suffolk and Nassau Counties* (New York: Hippocrene Books, 1991), 393.

"Council Rock (Oyster Bay, New York)." http://dictionary.sensagent.com/Council_Rock_(Oyster_Bay,_New_York)/en-en. This website contains history about George Fox.

Mary K. Peters, "Address at the Unveiling of a Tablet to George Fox at Council Rock" *Nassau County Historical Journal* 5, no. 1 (March 1942): 26–27. Background information is provided on George Fox.

Shelter Rock 70

Type of Formation: Large Boulder
WOW Factor: 5
Location: Manhasset (Nassau County)
Tenth Edition, NYS Atlas & Gazetteer: p. 111, C8
Earlier Edition, NYS Atlas & Gazetteer: p. 25, B6
Shelter Rock GPS Coordinates: 40°47.350′N, 73°41.463′W (estimated)
Accessibility: Roadside. A limited view of the boulder can be obtained through a chain-link fence; otherwise, the rock is inaccessible.
Degree of Difficulty: Easy

Description

Shelter Rock, aka Manhasset Rock and Milestone Rock, is a massive, granite boulder, with a 30-foot overhang, reputedly weighing up to 1,800 tons. It is roughly 55 feet by 35 feet.

History

A roadside sign near the boulder states that Shelter Rock was used for shelter by Native Americans (this was particularly true of the Matinecock tribe), as far back as 1,000 BC.

The site was excavated in 1946 by Carlyle S. Smith and Ralph Solecki for the American Museum of Natural History. Numerous artifacts were found, indicating a long-standing Native American presence.

In an article that appeared in the *Manhasset Press*, Steve Mosco states that for a period of time following World War I, Shelter Rock was the most photographed rock in the United States. Tales grew up about the rock—one even hinting that Captain Kidd may have hidden some treasure by the rock.

Today, the boulder is on land managed by the Greentree Foundation—a continuation of the 438-acre John Hay Whitney family estate founded in 1903. Whitney was a publisher and also served as ambassador to England.

Directions

From Manhasset (junction of Route 25A and Shelter Rock Road), drive south on Shelter Rock Road for 0.3 mile until you come to an intersection. Look for a historical marker describing the rock on the right side of the road virtually across from Old Shelter Rock Road. Here, next to the sign, can be seen the boulder, seemingly imprisoned behind a chain-link fence, 10 feet from the road.

The fact that the boulder lies close to what became the border between Manhasset and North Hills is not accidental. Like many big rocks in its day, Shelter Rock proved to be a convenient boundary line marker.

It should be noted that the Greentree Foundation occasionally leads tours to the boulder for special groups.

Resources

"So This Is the Longest Boulder on Long Island." https://www.scoutingny.com/stumbling-on-the-largest-boulder-on-long-island.

John Rather, "Shelter Rock Facing an Uncertain Future," *New York Times*, February 28, 1999," https://nytimes.com/1999/02/28/nyregion/shelter-rock-facing-an-uncertain-future.html.

"Shelter Rock Suffolk Gem and Mineral Club." https://www.suffolkgem.com/shelter
-rock.html.

Steve Mosco, "What Is Shelter Rock?" *Manhasset Press*, September 20, 2015. Website version is at https://manhassetpress.com/what-is-shelter-rock.

Carlyle S. Smith, "Manhasset Rock" (unpublished manuscript on file with the Department of Anthropology Archives at the American Museum of Natural History, June 10, 1946).

Margaret M. Voelbel, *The Story of an Island: The Geology and Geography of Long Island* (Point Washington, NY: Ira J. Friedman, 1965), 39. A drawing of the rock can be seen on page 39. The author writes, "there is a tremendous rock lying near the edge of the road . . . Years ago, when the road was a narrow dirt path, farmers driving cows along it would seek shelter under the overhang part of the rocks during storms. It has also been said that Indians used it for protection."

Big Rock and Little Rock 71

Type of Formation: Large Boulder
WOW Factor: 7–8
Location: Plandome Heights (Nassau County)
Tenth Edition, NYS Atlas & Gazetteer: p. 111, D8
Earlier Edition, NYS Atlas & Gazetteer: p. 25, B6
Junction of The Terrace & Plandome Road GPS Coordinates: 40°48.280′N, 73°42.374′W
Big Rock and Little Rock GPS Coordinates: 40°48.315′N, 73°42.465′W (general area)
Accessibility: Roadside or near roadside.
Degree of Difficulty: Unknown

Description

In Arlene Hinkemeyer's *A History of the Incorporated Village of Plandome Heights*, Big Rock is described as "a large glacial deposit up on the hill above what was then the end of Shore Road." In an old photograph, the rock appears to be 20–25 feet high. Hinkemeyer writes, "We would walk down The Terrace from Plandome Rd, and when the road made a right turn there was a path that went into the woods and right to Big Rock, and right across from it was another large rock called 'Little Rock.'" The woods are essentially gone, replaced by entire neighborhoods today.

History

The name Plandome is from the Latin words *planus* (plain) and *domus* (home) or "plain home."

Another rock was discovered in the Plandome/Manhasset area, although we suspect that it no longer exists. Margaret M. Voelbel, in *The Story of an Island: The Geology and Geography of Long Island*, writes, "When some new stores were being built on Plandome Road (in Manhasset) the workmen dug up a boulder that weighed 168 tons. It was just like the rock in the Palisades of New Jersey."

Directions

From Manhasset (junction of Plandome Road & Northern Boulevard), drive northwest on Plandome Road for 1.0 mile. Turn left onto The Terrace, a 0.2-mile-long road in the general area where the rocks are located. Likely, the path that led off from the road is gone, undoubtedly obliterated by the construction of new houses. You may have to ask around for the exact location of the rock since the area seems to be very residential.

Resources

Arlene Hinkemeyer, *A History of the Incorporated Village of Plandome Heights* (Manhasset, NY: Incorporated Village of Plandome Heights, 1997). A photograph of Big Rock is shown on page 8.

Margaret M. Voelbel. *The Story of an Island: The Geology and Geography of Long Island* (Port Washington, NY: Ira J. Friedman, 1965), 37.

Caumsett State Historic Park Boulders 72

Caumsett State Historic Park Preserve

Type of Formation: Large Boulder
WOW Factor: 4–5
Location: Lloyd Neck (Suffolk County)
Tenth Edition, NYS Atlas & Gazetteer: p. 112, A1
Earlier Edition, NYS Atlas & Gazetteer: p. 26, A1
Parking GPS Coordinates: 40°55.046′N, 73°28.377′W
Seashore Boulder GPS Coordinates: 40°56.332′N, 73°28.131′W
Fee: Admission charged
Hours & Days: Check with website for specific details
Accessibility: 1.8-mile walk to shoreline
Degree of Difficulty: *Walking*: Moderately difficult due to length of walk; *Biking*: Easy
Additional Information: Caumsett State Historic Park Preserve, 25 Lloyd Harbor Road, Huntington, NY 11743
Trail Map: parks.ny.gov/documents/parks/CaumsettTrailMap.pdf
The most efficient way of exploring the park would seem to be by bike.

Description

A number of medium-sized to large-sized boulders can be seen along the shoreline.

History

The site is the former 1,750-acre estate of Marshall Field III that Field purchased in 1921. The property was acquired by New York State in 1961.

Caumsett is Matinecock for "place by a sharp rock."

The Caumsett State Historic Park Preserve is located on Lloyd Neck, a small body of land 3.5 miles long by 2.0 miles wide and connected to the mainland by only the tiniest sliver of land.

Directions

From Huntington (junction of West Neck Road & Route 25A/Main Street), take West Neck Road northwest for 3.5 miles. At the end of the causeway between the mainland and Lloyd Neck, head east on Lloyd Harbor Road for 0.7 mile and then turn left into the entrance to Caumsett State Historic Park Preserve. Drive north for 0.2 mile to the parking area, which is to the right of the contact station.

Walk north along Fishing Drive Road for 1.8 miles to reach the fisherman's parking area and the shoreline. Only permit-holders are allowed to park in the fisherman's parking area.

From the parking area, walk to the beach where a couple of medium-sized boulders can be seen directly in front of the fisherman's parking area. From here, head southeast for <0.5 mile to reach the main boulder. Along the way, impressive, 100-foot-high bluffs are passed.

Resources

C. R. Roseberry, *From Niagara to Montauk: The Scenic Pleasures of New York State* (Albany, NY: State University of New York Press, 1982), 299–303.

New York-New Jersey Trail Conference, *Day Walker: 32 Hikes in the New York Metropolitan Area*, 2nd ed. (Mahwah, NJ: New York-New Jersey Trail Conference, 2002), 83.

"Caumsett State Historic Park Preserve." https://www.parks.ny.gov/parks/23/details.aspx.

Target Rock 73

Target Rock National Wildlife Refuge

Type of Formation: Large Boulder
WOW Factor: 5
Location: Huntington (Nassau County)
Tenth Edition, NYS Atlas & Gazetteer: p. 112, A1
Earlier Edition, NYS Atlas & Gazetteer: p. 26, B1
Parking GPS Coordinates: 40°55.630′N, 73°26.303′W
Target Rock GPS Coordinates: 40°55.778′N, 73°25.822′W
Fee: Modest admission charged
Hours & Days: Consult website for specific details
Accessibility: 0.4-mile hike
Degree of Difficulty: Moderate
Additional Information: Target Rock National Wildlife Refuge, Target Rock Road, Lloyd Neck, Huntington, NY; (516) 271-2409
Special use permits are required for any activity that requires accessing closed areas of the refuge, commercial activities, research and other miscellaneous events.
Map of Refuge: fws.gov/media/target-rock-trail-mappdf

Description

Target Rock is a 14-foot-high boulder jutting out of Huntington Bay whose flat sidewall proved to be an inviting target for sharpshooters.

History

The 80-acre Target Rock National Wildlife Refuge was donated in 1967 by the family of Ferdinand and Mary Eberstadt under the Migratory Bird Conservation Act. It is managed by the U.S. Fish and Wildlife Service.

29. Target Rock. Antique postcard, public domain.

The preserve and the rock's unusual name comes from the belief (probably true) that the British Navy used the rock for target practice during the Revolutionary War and War of 1812.

Directions

From Huntington (junction of West Neck Road & Route 25A/Main Street), take West Neck Road northwest for 3.5 miles. At the end of the causeway between the mainland and Lloyd Neck, head east on Lloyd Harbor Road for 2.7 miles. As the road turns left it becomes Target Rock Road. Continue north for 0.3 mile. As soon as you pass by Hawk Drive (to your right), you will come to the wildlife refuge's entrance, also on the right.

From the parking area, walk northeast for 0.4 mile to reach the beach, staying on the Rocky Beach Trail. Follow the shoreline north for less than 0.1 mile. Target Rock is directly offshore at East Fort Point, 150 feet from the beach.

Another, smaller, offshore boulder can be seen 0.1 mile southeast from Target Rock at 40°55.711′N, 73°25.797′W, as well as a variety of smaller boulders scattered along the beach.

Resources

"Target Rock National Wildlife Refuge." https://www.fws.gov/refuge/target-rock.
Raymond E. Spinzia, Judith A. Spinzia, and Kathryn E. Spinzia, *Long Island: A Guide to New York's Suffolk and Nassau Counties* (New York: Hippocrene Books, 1991), 118.
New York-New Jersey Trail Conference, *Day Walker: 32 Hikes in the New York Metropolitan Area*, 2nd ed. (Mahwah, NJ: New York-New Jersey Trail Conference, 2002), 79.
"Target Rock National Wildlife Refuge." https://ecos.fws.gov/ServCat/DownloadFile/38381.

Manhattan

Manhattan is a 12.5-mile-long, 2.5-mile-wide (at its widest) island that is delineated by the Hudson River to the west; the Spuyten Duyvil Creek to the north; the Harlem River, Hell Gate, and East River (really an arm of the Atlantic Ocean) to the east; and the East River and Upper New York Bay to the south. Perhaps not so strangely, 14 percent of Manhattan is built on a landfill.

The name Manhattan is a Native American Lenape word, meaning "island of many hills." The island's main natural feature is Central Park.

Central Park Rocks

74

Central Park

Type of Formation: Large Rock Mound; Medium-to-Large-Sized Boulder; Historic Cave
WOW Factor: 5
Location: Central Park (Manhattan)
Tenth Edition, NYS Atlas & Gazetteer: p. 110, D5
Earlier Edition, NYS Atlas & Gazetteer: p. 24, B1–2
Destination GPS Coordinates: *Umpire Rock*: 40°46.149′N, 73°58.672′W; *Cat Rock*: 40°46.099′N, 73°58.417′W; *Tooth Rock*: 40°46.020′N, 73°58.440′W; *Split Rock*: 40°46.563′N, 73°58.115′W; *Arch Rock* 40°15.706′N, 73°58.277′W; *Blockhouse Rock*: 40°47.890′N, 73°57.382′W (guesstimate) or 40°47.939′N, 73°57.367′W (Mountain Project); *Worthless Rock*: 40°47.898′N, 73°57.271′W; *Sheep Meadow Boulders*: 40°46.235′N, 73°58.471′W (primary); 40°46.380′N, 73°58.552′W (secondary #1); 40°46.302′N, 73°58.563′W (secondary #2)
Accessibility: Variable depending upon where you enter the park, and which rocks you are visiting
Degree of Difficulty: Easy to moderately easy
Central Park Maps: assets.centralparknyc.org/pdfs/maps/Central_Park_Map.pdf centralpark.com/locations/umpire-rock shows the location of Umpire Rock.

Description

Umpire Rock, aka Rat Rock, is a stupendous mound of exposed schist bedrock. It stands approximately 15 feet high and extends outward to form a crude circle approximately 55 feet wide. The rock has been a favorite gathering place for crowds of people for centuries. Like Drip Rock in another area of New York City, it is a major outcropping of bedrock.

Close by, to the north and east of Wollman Rink, are Cat Rock (perhaps named in response to Rat Rock), Chess Rock (named for its proximity to

30. Central Park Cave. Antique postcard, public domain.

the Chess and Checkers House at 40°46.140′N, 73°58.550′W), and Tooth Rock. By coincidence, there also happens to be a Cats Rock in Yonkers between Sprain Road and Grassy Sprain Road. However, it is more of a bluff than a defined rock and, more to the point, is not located in Central Park.

Blockhouse is a rock of significant size, which the Mountain Project describes as a 40–50-foot high slab.

Split Rock is a large boulder that has split into two halves with an appreciable gap between the two sections.

Worthless Rock is a 12–15-foot-high boulder near the roadside. A small slab has fractured off from the main body of the rock and lies close by.

Sheep Meadow Boulder is a large glacial erratic associated with the 33-acre Sheep Meadow, its namesake. A photograph of the rock can be seen in *The Conservationist* on page 13. Several medium-sized boulders resting on mounds of bedrock are also found in Sheep Meadow.

Arch Rock is located next to a 13-foot-high stone arch that helps form a 9-foot-high passageway. We don't have any description of the rock's size or shape.

Ramble Cave: During the early nineteenth century, a flight of steps led down from the stone arch mentioned above to an artificially created underground cave. Unfortunately, the cave was closed up in the 1920s, a victim of unintended consequences. According to M. M. Graff in *Central Park, Prospect Park: A New Perspective*, "The cave [was] sealed at both ends because of misuse by tramps. However, you can reach the steps, chiseled out of solid rock, by going south over the bridge and walking a few paces to the right; the treads pitch downward . . . until you stand under the immense looming rock that juts like a giant visor over the former boat landing."

History

Bill Thomas and Phyllis Thomas, in *Natural New York*, write, "The Park offers much to those interested in rocks and geology. Boulders more than a billion years old are found here, as well as some of the finest ice-polished rock anywhere. Layers of bedrock under the Park, known as the Manhattan foundation, were formed 480 million years ago. It's that same rock that makes it possible to build huge skyscrapers without their foundations giving way."

According to C. R. Roseberry in *From Niagara to Montauk: The Scenic Pleasures of New York State*, most of the boulders came from the Hudson Highlands and Palisades.

For more details about bouldering and rocks in this area, consult the *NYC Bouldering Guide* by Gareth "Gaz" Leah, and *A Climber's Guide to Popular Manhattan Boulder Problems* by Nicholas Falacci.

One of Central Park's glacial erratics made it into the 2008 science-fiction thriller movie *Jumper*. You will see the big Central Park glacial erratic near the beginning of the film.

Central Park was designed by Frederick Law Olmsted and Calvert Vaux. It remains a byproduct of the Romantic Revolution which, in turn, was a reaction to the Industrial Revolution and its despoiling of nature. The 840-acre park is rectangular in shape, 2.5 miles long, and over 0.5 mile wide. It contains 57 miles of pedestrian paths.

The northerly section of the park at the Harlem end is considerably more rugged than the southern section, rising up in places to as much as 100 feet above the rest of the landscape. The park is big!

UMPIRE ROCK

Umpire Rock was named for its close proximity to the Heckscher Ballfields, northwest of Umpire Rock's expansive, rocky mound. Umpire Rock's alternate name, Rat Rock, came about from swarms of rats that would emerge at night and engulf the rock. We believe this nocturnal occupation is no longer taking place.

BLOCKHOUSE ROCK

Blockhouse Rock is named for its location near the Blockhouse, a historic fortress that overlooks the flat surrounding areas north of Central Park. The fort was constructed in 1814 and used for defensive purposes. It is the second oldest structure in the park.

SHEEP MEADOW BOULDER

This boulder and others nearby are named for their proximity to Sheep Meadow. Initially, this tract of land was meant to serve as a parade ground. In 1864, however, it was put to more practical use when 200 sheep were brought into the park and housed in a Victorian building on the meadow. This practice lasted until 1934, when the sheep were relocated to Prospect Park in Brooklyn, and then later to the Catskills.

In more recent times, Sheep Meadow has been used for outdoor concerts and protests.

ARCH ROCK

Arch Rock is named for its location next to the Ramble Arch, east of the northeast end of the lake.

RAMBLE CAVE

Ramble Cave, aka Indian Cave, is another rock feature in Central Park, or should we say *was*, for it no longer exists. The cave (an artificial chamber) was created by the careful placement of large rocks to add diversity to the landscape. This was done after the original Olmsted and Vaux "Greensward Plan" was implemented. Unfortunately, the enclosure began to attract unsavory characters and, as a result, was sealed up in the 1930s. Today, only the steps leading down to the former entrance are visible, and they are hard to find unless you know exactly where to look.

Directions

Central Park occupies a considerable portion of the lower half of Manhattan delineated by Central Park South to the south, Central Park West to the west, Fifth Avenue to the east, and Central Park North to the north.

It is probably best to consult a map to guide you to Central Park. Our guess is that your initial approach would be from either the Henry Hudson Parkway (west side of Manhattan) or the Franklin D. Roosevelt East River Drive (along the east side of Manhattan).

UMPIRE ROCK

Umpire Rock is located near the southwest end of Central Park. From Central Park West and West 63rd Street, walk into the park and head southeast towards the south side of the Heckscher Ballfields. You will see the huge mound of bedrock between the Heckscher Ballfields and the Heckscher Playground.

CAT ROCK

Cat Rock is just a short distance northeast of the Victorian Gardens Amusement Park and Wollman Rink Skating School.

SPLIT ROCK

Split Rock is 200 feet northeast of the east end of the lake. From the Boathouse Restaurant, follow a path west to the Rambles. Look for a large boulder up on a small slope next to some benches.

BLOCKHOUSE BOULDER

Blockhouse Boulder is located not far from the north end of the North Woods section of Central Park, along a path behind the Block House.

WORTHLESS ROCK

Worthless Rock is 0.1 mile northwest of Harlem Meer, a small pond, 200 feet from West 110th Street. "Meer" is a Dutch word for "small sea." The boulder lies next to a paved walking path that parallels the road.

TOOTH ROCK

Tooth Rock can be found near the northeast end of "The Pond."

SHEEP MEADOW BOULDERS

Sheep Meadow Boulders are located in the large, grassy Sheep Meadow Field, not far from the south end of Central Park

Resources

Christopher J. Schuberth, *The Geology of New York City and Environs* (Garden City, NY: Natural History Press, 1968). On page 188 is a photograph of an unidentified glacial erratic in Central Park taken by Schuberth that is believed to have come from the Hudson Highlands.

"From Rat to Umpire: A Central Park Journey." https://amlit.commons.gc.cuny.edu/archives/574.

"The Ramble Cave – New York, New York – Atlas Obscura." https://www.atlasobscura.com/places/the-ramble-cave.
"Untapped Secrets of Central Park: Waterfalls, Caves, and Prehistoric Rocks." https://untappedcities.com/2013/05/20/secrets-central-park-waterfalls-caves-prehistoric-rocks.
"There Is a Hidden Cave in the Middle of Central Park." https://viewing.nyc/there-is-a-hidden-cave-in-the-middle-of-central-parks-ramble.
"Central Park and the Long-Lost Ramble Cave." https://bookwormhistory.com/2015/10/01/central-park-and-the-long-lost-ramble-cave.
Bill and Phyllis Thomas, *Natural New York* (New York: Holt, Rinehart and Winston, 1983), 154.
James Freund, *Central Park: A Photographic Excursion* (New York: Fordham University Press, 2001). On page 62 is an unidentified photograph of a huge rock mound, presumably Umpire Rock. On page 99 is a photograph of jumbo-sized rocks with the Dakota (a cooperative apartment building) in the background. This, then, would place this grouping of unnamed rocks near the intersection of Central Park West and 72nd Street.
Edward J. Levine, *Central Park*, Postcard History Series (Charleston, SC: Arcadia Publishing, 2006), 43. A photograph of the Rambles cave is shown before it was sealed up in 1929.
M. M. Graff, *Central Park, Prospect Park: A New Perspective* (New York: Greensward Foundation, 1985), 97–98.
"A Special Section: Central Park," *The Conservationist* 28, no. 4 (February–March 1974): 8–29.
"Geology of Central Park." www.geo.hunter.cuny.edu/research/archive/tcarboni_2016.pdf.
C. R. Roseberry, *From Niagara to Montauk: The Scenic Pleasures of New York State* (Albany, NY: State University of New York Press, 1982), 265.
Roy Rosenzweig and Elizabeth Blackman, *The Park and the People: A History of Central Park* (Ithaca, NY: Cornell University Press, 1992). The authors include an 1870 map of Central Park (pages 206–207) and then a modern map of the park (pages 464–465), revealing interesting changes.
Christopher Gray, "Central Park Indian Cave," *Northeastern Caver* 42, no. 3 (September 2011): 89. This is a piece taken from an article that appeared in the May 26, 2011 issue of the *New York Times*.
Lee Ann Levinson, *East Side, West Side: A Guide to New York City Parks in All Five Boroughs* (Darien, CT: Two Bytes, 1997), 81–91.
Elizabeth Barlow Rogers et al., *Rebuilding Central Park: A Management and Restoration Plan* (Cambridge, MA: MIT Press, 1987). A photograph of Umpire Rock is shown on pages 45 and 143.

Flood Rock (Historic) 75

Type of Formation: Destroyed Rock Island
WOW Factor: Unknown
Location: East River (Manhattan)
Tenth Edition, NYS Atlas & Gazetteer: p. 110, D5
Earlier Edition, NYS Atlas & Gazetteer: p. 24, BC4
Flood Rock former GPS Coordinates: 40°46.769′N, 73°56.246′W

31. Flood Rock explosion. Charles Graham, *The Last of Flood Rock—Scene of the Hell Gate Explosion, Viewed from 87th Street, NY*, 1885, wood engraving on paper, The Clark Art Institute, Williamstown, MA, 1955.4277. Image courtesy Clark Art Institute. clarkart.edu.

Description

Flood Rock is not really a rock, but rather a tiny, rocky island, nor can it be seen today, for it is long gone. We bring it to your attention because the largest, man-made, non-nuclear explosion in the world up to that time took place on the island.

Flood Rock Island, aka Mill Island and Great Mill Rock, and a smaller island called Little Mill Rock, were located in Hell Gate in the East River, and contributed greatly to Hell Gate's notorious reputation for being a deadly channel for ships to navigate through. It was for this reason that the rocky island was blasted apart.

History

Flood Rock (Island) was originally called Mill Rock (Island) after John Marsh erected a saw mill on the island in 1701.

Over a century later, in anticipation of the upcoming War of 1812, a U.S. Army fort was constructed on Mill Rock (Island), complete with a blockhouse and two cannons.

When both the War of 1812 and the Civil War had ended, government officials determined that the rocky island, now called Flood Island, no longer served any purpose other than being a navigational hazard. Hundreds of ships had been damaged, lost, or run aground trying to get through Hell Gate, the threat increased manifold by the presence of a giant whirlpool that frequently formed at the gateway. The U.S. Army Corps of Engineers were assigned the task of destroying the island. To accomplish this mission, they dug a tunnel under the East River from Queens to beneath Flood Rock. A subterranean chamber was loaded with about 300,000 pounds of high explosives and then detonated on October 10, 1885.

Flood Rock was instantaneously obliterated. The explosion was awesome and terrifying for anyone within earshot or eyeshot. What remained of the rubble was used to fill in around Little Mill Island, which became Mill Rock Park today.

Mill Rock Park has not been open to the public since the 1960s, despite a dock being present on the island's southern shore. It is presently home to nesting colonies of various birds.

Directions

Although Flood Rock no longer exists, the spot it occupied is about 0.1 mile southeast of Mill Rock Park.

The general area can be easily seen from 255-acre Ward Island Park (40°46.991′N, 73°56.100′W), northeast of Mill Rock; Hellgate Field, aka Astoria Athletic Field (40°46.655′N, 73°56.078′W), southwest of Mill Rock; and 15-acre Carl Schurz Park (40°46.614′N, 73°56.537′W), southwest of Mill Park.

Resources

Henry Collins Brown, ed., *Valentine's Manual of Old New York* (New York: Valentine's Manuel, 1927). On page 307 is a photograph of Flood Rock being blown up.

"Flood Rock (Former Site of Flood Rock/Great Mill Island)." http://wikimapia.org/9312630/Flood-Rock-Former-Site-of-Flood-Rock-Great-Mill-Island.

"The Blasting of Flood Rock." https://nygeschichte.blogspot.com/2010/06/blasting-of-flood-rock-1885.html.

Stonefield 76

Hudson River Park: Stonefield

Type of Formation: Large Rock
WOW Factor: 3
Location: Chelsea (Manhattan)
Tenth Edition, NYS Atlas & Gazetteer: p. 110, DE4
Earlier Edition, NYS Atlas & Gazetteer: p. 24, BC3–4
Parking GPS Coordinates: Park where a space is available along the streets
Stonefield GPS Coordinates: 40°45.009′N, 74°00.539′W
Accessibility: Between Pier 64 to north and Pier 62 to south
Degree of Difficulty: Easy
Additional Information: Chelsea Waterside Park, 185 Eleventh Avenue, New York City

---

Description

A number of large, quarried rocks have been placed in Stonefield for artistic purposes, the tallest stone being at least 10 feet in height.

History

Stonefield is part of the 550-acre Hudson River Park that includes the Chelsea Waterside Park.

The stones used for Stonefield were taken from quarries in New York State and northeast Pennsylvania, and chosen for their shape, size, and color. The rock-strewn landscape was created by artist Meg Webster, a postminimalist known for her sculptures and installation art. Piers 62, 63, and 64 were designed by Michael Van Valkenburgh, an American landscape architect and educator.

Stonefield was made possible through the Hudson River Park Trust, a state-created entity that oversees the planning, design, construction, and maintenance of the waterfront park.

Chelsea was named by a British major, Thomas Clarke, who brought the land in 1750. Interestingly, Clarke was the grandfather of Clement Clarke Moore, author of the famous poem "A Visit from St. Nicholas."

Historically, the Chelsea piers were used extensively as ports for ocean liners, particularly during the 1920s up to the mid-1940s.

Directions

The rocks are part of the Chelsea Waterside Park near the junction of Eleventh Avenue and West 23rd Street.

Park wherever you can find a space—this is Manhattan after all.

Resources

"Stonefield Landscape Sculpture." https://hudsonriverpark.org/activities/stonefield-native-garden/.

Sanna Feirstein, *Naming New York Manhattan Places and How They Got Their Names* (New York: New York University Press, 2001), 95.

Stanley Wilcox and H. W. Van Loan, *The Hudson from Troy to the Battery* (Philmont, NY: Riverview Publishing, 2011), 138.

Point of Rocks

77

St. Nicholas Park

Type of Formation: Historic Rocky Bluff
WOW Factor: 3
Location: Central Harlem (Manhattan)
Tenth Edition, NYS Atlas & Gazetteer: p. 110, D5
Earlier Edition, NYS Atlas & Gazetteer: p. 24, AB2
St. Nicholas Park GPS Coordinates: 40°49.011′N, 73°56.950′W
Point of Rocks GPS Coordinates: 40°48.776′N, 73°57.059′W (estimated)
Accessibility: Roadside
Degree of Difficulty: Easy

32. Point of Rocks. *Source*: Carl Horton Pierce, *New Harlem Past and Present* (New York: New Harlem Publishing, 1903), 258–259 insert.

Description

Point of Rocks is a small rocky bluff, fairly undistinguished except for its history. We don't know for sure if the rocky bluff still exists in its original form. Possibly, it has been modified structurally over the last two hundred years.

History

It was while standing on top of the Point of Rocks that General George Washington successfully directed his forces to outmaneuver Lord Howe's British forces at the Battle of Harlem Heights.

The 23-acre St. Nicholas Park, a "ribbon park" due to its narrow length, was designed by Samuel Parsons Jr. to preserve "the rugged qualities of the natural landscape" and was created in 1895. Additional land was acquired over the next decade, including the property from West 130th Street to West 128th Street that included Point of Rocks.

The park's name was taken from two adjacent streets, St. Nicholas Terrace to the west and St. Nicholas Avenue to the east.

Directions

St. Nicholas Park extends from West 128th Street to West 141st Street, about midway between the Henry Hudson Parkway and the Harlem River Drive, roughly seventeen blocks north of Central Park. Because there's no simple way to describe how get to the park, we would suggest using a map of Manhattan or MapQuest as your guide.

Once you have reached St. Nicholas Park, look for the Point of Rocks near the southeastern corner of the park by the junction of St. Nicholas Avenue and 129th Street.

Resources

William Pennington Toler and Harmon De Pau Nutting, *New Harlem Past and Present* (New York: New Harlem Publishing, 1903).

"St. Nicholas Park." https://www.nycgovparks.org/parks/st-nicholas-park/history.

Erik K. Washington, *Manhattanville: Old Heart of West Harlem*. Images of America Series (Charleston, SC: Arcadia Publishing, 2002). A photograph of Point of Rocks is shown on page 74.

Andrew S. Dolkart and Gretchen S. Sorin, *Touring Historic Harlem: Four Walks in Northern Manhattan* (New York: New York Landmark Conservancy, 1997), 97. A photograph of the park is on page 98.

Carl Horton Pierce, *New Harlem: Past and Present* (New York: New Harlem Publishing, 1903. A photo of the rocky bluff is in an insert between pages 258 and 259.

Queens County

Queens is the easternmost and largest of the five boroughs of New York City. Its northern portion consists of a rolling landscape, but with no hills of any great height. The boulders found are mostly granite.

In the southern portion of the borough, virtually no rocks of any significant size are likely to be encountered.

Giant Rock

78

Type of Formation: Large Boulder
WOW Factor: 7
Location: East Elmhurst (Queens County)
Tenth Edition, NYS Atlas & Gazetteer: p. 111, D6
Earlier Edition, NYS Atlas & Gazetteer: p. 25, BC4–5
Parking GPS Coordinates: 40°46.074′N, 73°52.039′W (estimated)
Giant Rock GPS Coordinates: 40°46.068′N, 73°52.047′W
Accessibility: Roadside
Degree of Difficulty: Easy

33. Giant Rock. Photograph by the author.

Description

The Giant Rock, aka Big Boulder, Pet Rock, and Ditmars Boulevard Crowne Plaza Pet Rock is a 1,000-ton, 6- to 10-foot-high boulder primarily made up of granite pegmatite.

History

One writer has said that if there were a New York City version of the Seven Wonders of the World, Giant Rock would be on the list.

Some years ago, a competition was held between hotel employees at the Crowne Plaza Hotel and the Hampton Inn to come up with a name for the rock. "Pet Rock" won out, reflecting the fact that local children had been burying their pets near the humongous boulder for years.

Upon hearing about the result of this contest, however, the Corona-East Elmhurst Historical Society promptly notified both hotels that the boulder already had an official name, and that name was Giant Rock. So Giant Rock it is and forever will be.

Although Giant Rock has been at its current location for 10,000–12,000 years and seemed like a permanent feature of the landscape, it was almost destroyed by overeager developers. In 1980, the Crowne Plaza Hotel wanted to obliterate the rock. Fortunately, irate citizens rose up to defend the rock and won out. Then, 19 years later, it was the Hampton Inn's turn to target the rock. Facing fierce opposition, they also relented. Giant Rock, however, bears scars from the initial attempts that were made to remove it.

Today, the rock displays a plaque on it that reads (in part), "This 1000-ton boulder was brought to its present location (probably from southern Westchester) by an ice sheet about 10,000 or 12,000 years ago. Although the boulder is impressive, it is only a small part of the ice sheet's load. Long Island is built almost entirely of materials (boulders, sand, gravel, and clay) that were brought here by ice."

Directions

The boulder is located between the Hampton Inn and Crowne Plaza Hotel, off Ditmars Boulevard, southwest of the Grand Central Parkway, and very close to LaGuardia Airport.

Precise directions to this general location can be obtained from the website of either hotel.

Resources

"LaGuardia Airport's 'Giant Rock' Is Older Than Human History." https://gothamist.com/arts-entertainment/laguardia-airports-giant-rock-is-older-than-human-history.

"About That Giant Rock." https://www.ipetitions.com/petition/about-that-giant-rock.

"The 'Giant Rock' Historic District." https://6tocelebrate.org/site/the-giant-rock. The website shows a photograph of the large rock.

White Stone 79

Type of Formation: Large Boulder
WOW Factor: 6
Location: Whitestone (Queens)
Tenth Edition, NYS Atlas & Gazetteer: p. 111, D7
Earlier Edition, NYS Atlas & Gazetteer: p. 25, B5
Francis Lewis Park **Parking GPS Coordinates:** 40°47.761′N, 73°49.451′W
White Stone/Rock Point GPS Coordinates: 40°48.059′N, 73°49.191′W
Accessibility: 0.4-mile paddle by water
Degree of Difficulty: Easy
Additional Information: In order to launch a kayak, canoe, or boat, it is necessary to first obtain a permit. For more information, consult nycgovparks.org/pagefiles/127/Kayak-Canoe-Boat-Launch-Permit_5ade12c252c6c.pdf.

Description

In "The Geology of Long Island," Jay T. Fox mentions a glacial boulder at Whitestone Landing that measures 20 feet in height. Presumably, this is White Stone, right on Whitestone Point.

Two sizeable boulders can be seen on Rock Point (the blunted, arrowhead-like projection of land at Whitestone), one of which may be White Stone.

History

According to Jason D. Antos in *Whitestone: Images of America*, it was the unusual white rock that gave Whitestone its name; however, the name may also be a reference to the white limestone upon which the town is built.

Directions

Going north on I-678/Bronx Whitestone Expressway, take Exit 17 for Third Avenue. Continue north on the Whitestone Expressway (a one-way, smaller version of the expressway) for 0.2 mile to reach, at the road's end, Francis Lewis Park (located at the junction of the Whitestone Expressway & Third Avenue). Park along either side of the street in front of the park.

WATER ACCESS

From Francis Lewis Park (named in honor of one of the signers of the Declaration of Independence), launch a kayak or canoe and paddle northeast for 0.4 mile (bypassing a number of private docks) to reach the rocks. You will need to apply for a permit to do so, however.

If you paddle out to see the boulders at Whitestone, make a point to continue over to the automated light on a tiny island less than 0.1 mile from the shore. Its entire perimeter is lined with massive rocks, making it seem almost like a rock fortress. This is what Google Earth labels as Whitestone Rock.

LAND ACCESS

To be sure, it might also be possible to reach the rocks from nearby Powells Cove Boulevard (northwest of Francis Lewis Park), but a wall of private homes between the street and the East River presents what may be an insurmountable barrier unless you happen to know one of the homeowners or get permission to cross a resident's yard in order to reach the shoreline.

Resources

Jason D. Antos, *Whitestone*, Images of America Series (Charleston, SC: Arcadia Publishing, 2006), 14.

"It's Called a Great Place to Live." https://www.qchron.com/editions/queenswide/it-s-called-a-great-place-to-live/article_4a3d2752-556e-550d-a969-ac5045dff5f8.html. This website mentions that the boulder is made of limestone and is located on the shore.

Long Island City Glacial Erratic 80

Type of Formation: Large Rock
WOW Factor: 2
Location: Hunters Point/Long Island City (Queens County)
Tenth Edition, NYS Atlas & Gazetteer: p. 119, E4
Earlier Edition, NYS Atlas & Gazetteer: p. 24, BC2
Glacial Erratic GPS Coordinates: 40°45.024′N, 73°56.888′W
Accessibility: Roadside
Degree of Difficulty: Easy

Description

This rock, said to be a glacial erratic, occupies a fair amount of space, but is not particularly high, possibly being no taller than the roof of an automobile. Personally, we would be more inclined to call it a mound of bedrock than a glacial erratic unless, that is, the rock is merely the top of a buried boulder. The rock is partitioned off by two guardrails.

History

This bulky rock, which for us remains nameless, partially blocks 12th Street, and yet has been obviously tolerated by the city, which has put up the street and buildings around the rock instead of blasting it apart. Automobiles also give the rock a wide berth.

Directions

Drive southeast across the East River via the Ed Koch Queensboro Bridge/Route 25 to reach the Hunters Point section of Long Island City. The rock is at the junction of 12th Street and 43rd Road, north of where Route 25/Queens Boulevard crosses over Route 25A/Jackson Avenue. You will need to consult a map or MapQuest to figure out the best way to get to the rock once you are at Hunters Point.

34. Long Island City Glacial Boulder. Photograph by the author.

What's unusual about this rock is that instead of being seen as a nuisance to progress, it has been integrated into the city's infrastructure.

Amazingly, this is not the only rock left untouched by developers in New York City. Two more rocks, both giants, are described in Chapter 152, section 55.

Resources

"Long Island City." https://everipedia.org/wiki/lang_en/Long_Island_City. This website contains the only photograph of the rock that we have seen.

Staten Island (Richmond County)

Staten Island is an oval-shaped island, 14 miles long and 8 miles wide at its maximum, encompassing 64 square miles of land. It is the southwesternmost of the five boroughs of New York City. The island is separated from the rest of New York City by New York Bay, and from New Jersey by the Arthur Kill, aka Staten Island Sound, and the 3.0-mile-long Kill Van Kull.

Staten Island was named by Henry Hudson in honor of the Dutch governing body that authorized his expedition. Technically, Native Americans had first rights on nomenclature. The Algonquian called it "Aquehonga Manachnong," meaning "as far as the place of the bad woods" and "Eghquhous," or "the bad woods." Is it possible that Native Americans held unfriendly feelings about these particular woods? Interestingly, the island was known as Richmond (a name imposed on it by the British) until 1975, when it officially became Staten Island.

The landscape is elevated and fairly broken up. According to J. H. Mather and L. P. Brochett, in *A Geographical History of the State of New York* (1848), "Boulders of green-stone, sandstone, gneiss, granite & etc., appear in some sections sparingly, but on the northeast part of the island in considerable abundance."

Moses Mountain

81

High Rock Park

Type of Formation: Rock Mound
WOW Factor: 5
Location: Richmond Town (Richmond County)
Tenth Edition, NYS Atlas & Gazetteer: p. 116, AB3
Earlier Edition, NYS Atlas & Gazetteer: p. 24, CD3
Parking GPS Coordinates: 40°34.988′N, 74°07.837′W
Moses Mountain GPS Coordinates: 40°35.080′N, 74°07.804′W
Accessibility: 0.3-mile hike
Degree of Difficulty: Moderately easy
Additional Information: Moses Mountain, 1770–1778 Manor Road, Staten Island, NY 10306

Description

What's both interesting and unique about 260-foot-high Moses Mountain, aka Mount Moses, is that it is an artificial hill made up of rock and gravel carted away from a nearby location being excavated at the time.

In many ways, it is reminiscent of prehistoric Silbury Hill—a towering, 130-foot-high mound that was created by Stone Age people—near Avebury, England.

Small-to-mid-sized rocks can be seen at the summit, called Paulo's Peak.

There is a quasi-lookout at the top, with a small rock that you can stand on to raise your height another two feet. From the summit you can see the Greenbelt area and New Jersey's Atlantic Highlands.

History

The history of this mountain goes back to the early 1960s when New York City planner and parks commissioner Robert Moses began to construct the Richmond Parkway over Todt Hill through what has become known as the Greenbelt, and gravel blasted away was piled up in a remote area that grew in height, becoming known, somewhat ironically, as Moses Mountain. The four-mile-long road was never completed, however. Environmentalists and wildlife conservators sued city officials, Governor Nelson Rockefeller, and Mayor John Lindsay to stop the project and the judge ruled in their favor.

After the Staten Island Greenbelt was formed in 2021, the mountain was renamed Paulo's Peak, for Thomas A. Paulo, former New York City parks borough commissioner for Staten Island.

Directions

Driving east on the I-278/Staten Island Expressway, take Exit 8 for Victory Boulevard, and follow the connecting ramp south for 0.3 mile. Turn right onto Victory Boulevard and go southwest for 0.4 mile. Turn left onto Richmond Avenue and proceed south for 0.8 mile.

Driving west on the Staten Island Expressway, take Exit 7 for Richmond Avenue. Turn right onto Richmond Avenue and head south for 1.5 miles.

From either approach, turn left onto Rockland Avenue and proceed southeast for 2.0 miles.

Park on the north side of Rockland Avenue, 100 feet northwest of its junction with Manor Road. Follow a path that leads northeast to the tiny summit in <0.3 mile.

Resources

"Blood Root Valley: Paulo's Peak." https:/www.nycgovparks.org/parks/the-greenbelt/highlights/19851.

"The Greenbelt: Paulo's Peak." https://www.nycgovparks.org/park-features/virtual-tours/greenbelt/moses-mountain.

Sugar Loaf Rock

82

Hero Park

Type of Formation: Large Boulder
WOW Factor: 3–4
Location: Richmond Town (Richmond County)
Tenth Edition, NYS Atlas & Gazetteer: p. 116, A3
Earlier Edition, NYS Atlas & Gazetteer: p. 24, CD3
Hero Park GPS Coordinates: 40°37.830N, 74°05.278′W
Sugar Loaf Rock GPS Coordinates: 40°37.813N, 74°05.249′W
Accessibility: <0.05-mile walk
Degree of Difficulty: Easy

35. Sugar Loaf. Antique postcard, public domain.

Description

According to Ira K. Morris, in volume 1 of *Morris's Memorial History of Staten Island, New York*, there is "a prominent boulder, the shape of a sugar loaf, near the paper factory at the corner of Prospect Street and the turnpike. It now occupies a clear field, but was once surrounded by woods, and was then a point of pilgrimage for the boys of the period."

History

During early years, Sugar Loaf Rock served as a camping and gathering place for Native Americans. In recent times, the rock became the centerpiece of Hero Park—a 1.1-acre tract of land that honors 144 Staten Island soldiers who died during World War I. A tablet stating, "This granite boulder left here during the Glacial Period has been known for generations as Sugar Loaf Rock and marks the boyhood playground of the men whose gallant deeds it now commemorates" was removed in the 1970s following acts of vandalism. In 2006, the park underwent restoration, and the dedication plaque was recreated.

The original tract of land, of which Hero's Park is but a fraction, was donated to the city in 1920 by Dr. Louis. A. Dreyfus (who is remembered for having discovered the process for making a chewing gum base in 1909).

Directions

From the I-278/Staten Island Expressway, take Exit 12 for Slosson Avenue & Todt Hill Road and head north on Slosson Avenue for <0.2 mile. Turn right onto Victory Boulevard and follow it northeast for 2.0 miles to reach Hero Park, on your right.

This park, bounded by Victory Boulevard to the west, Louis Street to the north, and Howard Avenue to the east, is tiny and readily accessible.

Resources

Ira K. Morris, *Morris's Memorial History of Staten Island, New York*, vol. 1 (New York: Memorial Publishing, 1898), 372.

Charles Gilbert Hine, comp., *History and Legend of Howard Avenue and the Serpentine Road, Grymes Hill, Staten Island* (printed by the author, 1914). Between pages 16 and 17 is a photograph of a large rock with the caption "Sugar Loaf of Druid's Rock."

"Hero Park." https://www.nycgovparks.org/parks/hero-park/history.

Stonehenges Shores 83

Mount Loretto Unique Area Nature Preserve

Type of Formation: Art Rock
WOW Factor: 3–4
Location: Tottenville (Richmond County)
Tenth Edition, NYS Atlas & Gazetteer: p. 116, B2
Earlier Edition, NYS Atlas & Gazetteer: p. 24, D2
Parking GPS Coordinates: 40°30.560′N, 74°13.090′W
Stonehenges Shores GPS Coordinates: 40°30.451′N, 74°12.753′W
Hours & Days: Daily, dawn to dusk
Accessibility: 0.7-mile walk
Degree of Difficulty: Easy
Additional Information: Mount Loretto Unique Area Nature Preserve, 6450 Hylan Boulevard, Staten Island, NY
Preserve Map: alltrails.com/trail/us/new-york/mount-loretto-grassland-and-wetland-loop
Mount Loretto Unique Area Map: dec.ny.gov/docs/regions_pdf/map mtloretto.pdf

---

Description

Chris Gethard, in *Weird New York*, writes that "one is suddenly transported to an alien landscape as hundreds of stone towers, roads, rooms, and driftwood monoliths rise from the beach" at Princess Bay. "The curious sculptures, which are composed of materials found along the high tide line, rise seemingly out of nowhere to meet you."

These driftwood monoliths and sculptures are the product of one man's creative imagination, Douglas Schwartz, who has taken rocks, debris, and other materials gathered along the high-tide line to create artistic works on the beach. Others have since joined in as well. Some have referred to the art project as the "rock sculptures of Staten Island."

The rocks are at the base of 75-foot-high red clay cliffs and fill a 0.5-mile length of beach.

History

Douglas Schwartz, by profession, was not only a zookeeper but also an artist. His cairn-like stone monoliths are found along the beach between Sharrott Avenue and Page Avenue.

From the late 1800s through most of the 1900s, Mount Loretto (from which the Mount Loretto Unique Area Nature Preserve takes its name) was the site of St. Elizabeth's Home for Girls. The property was purchased by New York State in 1999 following much advocacy from the Trust for Public Land and Protectors of Pine Oak Woods, and is now the 241-acre Mount Loretto Unique Area Nature Preserve. It is owned and managed by the New York State Department of Environmental Conservation (DEC).

The Prince's Bay Lighthouse, aka Red Bank Lighthouse and the John Cardinal O'Connor Lighthouse, was built in 1828 and then refurbished in 1864. It was deactivated in 1922. For many visitors, this is the primary destination.

Directions

Driving east on Route 440/Korean War Veterans Parkway, take Exit 1 for Page Avenue. Turn right onto Boscombe Avenue, which becomes Page Avenue as you bear left and head south for 1.2 miles to Hylan Boulevard.

Driving west on Route 440/Korean War Veterans Parkway, take Exit 1 for Arthur Kill Road. When you come to Veterans Road West, turn right and proceed east for 0.2 mile. At Tyrellan Avenue, turn right and head south for 0.2 mile. Then bear right onto Boscombe Avenue. Follow it west, then south as it turns into Page Avenue and reaches Hylan Boulevard in 1.5 mile.

From either approach, turn left onto Hylan Boulevard and proceed east for >0.6 mile and turn right into the parking area for the Mount Loretto Unique Area Nature Preserve by a large, open field.

From the parking area, walk past the barrier and follow Kenny Road south for 0.4 mile. At the end of the road, overlooking the shore, turn left and go past a small shrine. Until a fire in 2000, an orphanage dormitory and hospital for the Archdiocese of New York also stood here.

Follow the Beach Loop Trail and continue east along the shoreline for 0.3 mile to the start of the rock sculptures. Along the way, you will pass by the Prince's Bay Lighthouse.

Hopefully, the years have not taken their toll on Schwartz's works. Perhaps even new ones have sprung up in addition to the old ones. Still, one must always bear in mind that rock sculptures can be transitory, especially when subjected continuously to the vicissitudes of wind, rain, snow, and ice.

Resources

Chris Gethard, *Weird New York: Your Travel Guide to New York's Local Legends and Best Kept Secret* (New York: Sterling, 2005), 161.

Jim O'Grady, "Neighborhood Report: St. George; The Art That Washes In," September 20, 1998, https://nytimes.com/1998/09/20/nyregion/neighborhood-report-st-george-the-art-that-washes-in.html.

James Curcuru, "The Rock Artist," *Gelf Magazine*, June 30, 2008, www.gelfmagazine.com/archives/the_rock_artist.php.

"Mt. Loretto's Rock Sculptures Forgotten New York." https://forgotten-ny.com/2007/09/mt-lorettos-rock-sculptures.

"Mount Loretto Beach Rock Garden – Atlas Obscura" https://atlasobscura.com/places/mount-loretto-beach-rock-garden.

Field Guide to the Natural World of New York City (Baltimore, MD: Johns Hopkins University Press, 2007). A photograph of the rock sculptures can be seen on page 99.

"The Rock Sculptures of Douglas Schwartz." https://www.odditycentral.com/pics/the-mysterious-rock-sculptures-of-staten-island.html.

"Zookeeper Douglas Schwartz Pulls Staten Island Chuck." https://www.gettyimages.com/detail/news-photo/zookeeper-douglas-schwartz- pulls-staten-island-chuck-from-news-photo/97249095.

Fort Hill Park Boulder 84

Fort Hill Park

Type of Formation: Large Boulder
WOW Factor: Unknown
Location: North of Richmond Town (Richmond County)
Tenth Edition, NYS Atlas & Gazetteer: p. 116, A3–4
Earlier Edition, NYS Atlas & Gazetteer: p. 24, CD3
Parking GPS Coordinates: Park where you can along the adjacent streets
Fort Hill Park GPS Coordinates: 40°38.449′N, 74°04.893′W
Fort Hill Park Boulder GPS Coordinates: Not determined
Accessibility: 0.05-mile hike
Degree of Difficulty: Moderately easy
Additional Information: nycgovparks.org/fort-hill-park/map

Description

In *History of Richmond County, (Staten Island) New York, from Its Discovery to the Present Times*, mention is made that "one of these large boulders rests directly on top of Fort Hill, New Brighton." The park on Fort Hill is small, occupying <0.1-acres of land. Locating the boulder, then, should not prove difficult.

History

Fort Hill was once known as Fort Knyphausen, named for Baron von Knyphausen, a Prussian general who built the fort for the British during the American Revolution. It successfully repelled the American Continental Army in 1780. In 1783, the fort was abandoned, and its square base later used as a reservoir.

Fort Hill Park covers less than 0.9 acres, surrounded by a heavily populated area. It was acquired by the city and dedicated as a park in 2004. A bluestone sidewalk leads into the park.

Directions

Driving along the I-278/Staten Island Expressway, take Exit 11 for Bradley Avenue and proceed north on Bradley Avenue for 0.3 mile. Turn right onto Victory Boulevard and proceed northeast for 3.3 miles. Then bear left onto Westervelt Avenue and go north for >0.2 mile. When you come to Benziger Avenue, turn right, go east for 0.1 mile, and left onto Daniel Low Avenue (a one-way street). Head north for 0.1 mile and then left onto Fort Place. Going left will take you onto Fort Place, a one-way street with parking on the right. Park along the side of the street or turn right onto Hendricks Avenue (also a one-way street) and park. Fort Park is to your right.

We assume the park entrance is off of Fort Hill Road, which is near the intersection of Hendricks Avenue and Westervelt Avenue. One thing is for certain. The park is surrounded by many private homes. Finding the right approach to the park could be the main issue.

Resources

Richard M. Bayles, ed., *History of Richmond County, (Staten Island) New York, from Its Discovery to the Present Times* (New York: L. E. Preston, 1887), 15. Bayles also mentions that, besides at Fort Hill, "moderately large boulders, both of trap and gneiss, abound on the moraine between the Narrows and Garretson's."

"Fort Hill, New Brighton Forgotten New York." https//forgotten-ny.com/2018/02/fort-hill-new-brighton. This website contains a great deal of information about Fort Hill and the houses built around it.

Long Island

Long Island was appropriately named, for it is a very long body of land indeed (118 miles to be exact). It consists of a chain of low hills, remnants of a glacial, terminal moraine that divided the island into two distinctive sections. To the north, the landscape is rough and broken, its shoreline strewn with boulders. To the south, the island is a relatively flat plain.

If you look at Long Island dispassionately, it assumes the shape of a fish whose tail fins are located at the island's east end, the shorter fin being Orient Point and the longer one Montauk Point. Or, to use an even more colorful description borrowed from Christopher Bollen's novel *Orient*, "This is how we first saw you, Long Island . . . Like the body of a woman floating in New York harbor. It still amazes me that no one else sees the shape of a woman in that island separated along the coastline, her legs the two beach-lined forks that jut out to sea when the land splits, her hips and breasts the rocky inlets of oyster coves, her skull broken in the boroughs of New York City."

Suffolk County

Suffolk County is the easternmost county in New York State, located at the east end of Long Island, occupying 66 percent of Long Island's surface. The other section of Long Island belongs to Nassau County, part of which has been included in a previous section of this book.

Approaching the northern shore, the ground is hilly and broken. To the south, the landscape is level and sandy, virtually flat.

The county took its name from Suffolk, England.

Halesite Boulder

85

Type of Formation: Medium-Sized Boulder; Historic
WOW Factor: 3
Location: Halesite (Suffolk County)
Tenth Edition, NYS Atlas & Gazetteer: p. 112, B1–2
Earlier Edition, NYS Atlas & Gazetteer: p. 26, AB1
Parking GPS Coordinates: 40°53.075′N, 73°25.165′W
Halesite Boulder GPS Coordinates: 40°53.036′N, 73°25.163′W
Accessibility: Roadside; 250-foot walk from parking area
Degree of Difficulty: Easy

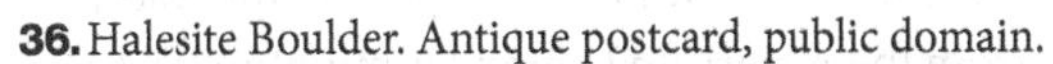

36. Halesite Boulder. Antique postcard, public domain.

Description

The Halesite Boulder, more commonly known as the Nathan Hale Monument, is a 5-foot-high boulder that previously overlooked Huntington Harbor.

The Halesite Boulder is not the only significant rock in the area. In "The Geology of Long Island," Jay T. Fox mentions a "20-foot glacial boulder, ½ mile south of Halesite," which would put the boulder somewhere halfway between Halesite and Huntington. We have not been able to pinpoint its location, however, hardly surprising given the surface area involved.

History

Nathan Hale was a colonial patriot and spy for the Continental Army who is famous for having said at the time of his hanging in 1776 (but perhaps not quite as poetically as told by legend) that "I only regret that I have but one life to lose for my country."

The Halesite Boulder originally rested on a nearby beach, where it overlooked the spot where Hale is thought to have landed while setting off to accomplish his mission. A plaque honoring Hale was subsequently mounted on the rock.

Later, the boulder was moved to the south end of the bay. Due to road construction around 2012, however, the boulder was again relocated, this time a distance of only seventy feet from its previous spot. It now rests permanently in place across from the American Legion Huntington Post 360.

Directions

From Huntington (junction of Routes 110/New York Avenue and 25A/Main Street), drive northeast on Route 110/New York Avenue for 1.0 mile. At the second roundabout, head west on Route 35/Mill Dam Road for 150 feet and turn right into the parking area for the American Legion Post.

From here, walk to the front of the American Legion Post and then south across the Mill Dam Road crosswalk. The rock is to your left.

Resources

"Nathan Hale Rock Moved to Accommodate Halesite Construction." https//thehuntingtonian.com/2012/09/17/nathan-hale-rock-moved-to-accommodate-halesite-construction.

"Nathan Hale Everlasting Revolutionary War Memorial." htps://www.nationalwarmemorialregistry.org/memorials/nathan-hale-everlasting-remembrance-revolutionary-war-memorial/.

Stony Brook University South Entrance Boulder

86

Stony Brook University

Type of Formation: Large Boulder
WOW Factor: 4–5
Location: Stony Brook University (Suffolk County)
Tenth Edition, NYS Atlas & Gazetteer: p. 112, B5
Earlier Edition, NYS Atlas & Gazetteer: p. 26, A3–4
South Entrance Boulder GPS Coordinates: 40°54.184′N, 73°06.979′W
Accessibility: Roadside
Degree of Difficulty: Easy

Description

This large boulder has been described as big as a pickup truck.

There is also another medium-sized boulder (the size of an automobile) that is next to one of the parking lots at Stony Brook University. We have not been able to find it using Google Earth but have seen a photograph of it.

History

A plaque in front of the boulder provides information about its geological history.

Stony Brook University was formed in 1957 to help secondary school teachers prepare to teach science and mathematics. The college was originally located in Oyster Bay, relocating to Stony Brook University in 1962 on land donated by philanthropist Ward Melville.

Directions

From north of Centereach (junction of Routes 97/Nicholls Road & 347/ Smithtown Bypass/Nesconset Highway), drive north on Route 97/Nicholls Road for 1.5 miles. Turn left onto South Drive and then, at the soonest possible moment, turn around, and return to Route 97/Nicholls Road, this time heading south. After 250 feet, pull over to your right at the end of the guardrails. You will see a large boulder on your right.

Resources

"Stony Brook University Boulder Climbing." https://mountainproject.com/area/116124752/stony-brook-university-boulder.

Patriot's Rock

87

Type of Formation: Large Rock
WOW Factor: 7
Location: East Setauket (Suffolk County)
Tenth Edition, NYS Atlas & Gazetteer: p. 112, A5
Earlier Edition, NYS Atlas & Gazetteer: p. 26, A3–4
Parking GPS Coordinates: 40°56.714N, 73°06.880'W
Patriot's Rock GPS Coordinates: 40°56.697'N, 73°06.822'W
Accessibility: 70-foot walk
Degree of Difficulty: Easy

Description

Patriot's Rock, aka The Rock, is located in a 3.5-acre park, and serves as the park's centerpiece. The rock is about 10 feet high and 25 feet long, according to the *Three Village Guidebook: The Setaukets, Poquott, Old Field & Stony Brook.*

History

Patriot's Rock's name arose from its role at the "Battle of Setauket" in 1777. The patriots, led by Brigadier General Samuel Holden Parsons, used the rock for cover while they fired a canon at British loyalists led by Lieutenant Colonel Richard Hewlett, who had taken refuge across the village green in the nearby Presbyterian Church.

Many years later, the plaque that adorns the rock was set into place by the Daughters of the American Revolution.

According to Raymond E. Spinzia, Judith A. Spinzia, and Kathryn E. Spinzia, in *Long Island: A Guide to New York's Suffolk and Nassau Counties*, the Reverend Nathaniel Brewster, who was Setauket's first ordained minister, preached his first sermon from the top of the rock in 1665.

A painting of Patriot's Rock by William Sidney Mount, called *The Rock on the Green*, depicts the boulder as it looked at the time of the Revolutionary War. It hangs in the Long Island Museum of Art at Stony Brook [40°54.696'N, 73°08.517'W].

Prior to European occupation, the rock was used by the Setalcott, an indigenous tribe, as a meeting place.

Directions

From Stony Brook (junction of Routes 25A/North Country Road & 97/ Nicholls Road), head northeast on Route 25A/North Country Road for >1.3 miles. Turn left onto Main Street and head northwest for 0.6 mile to reach Setauket Mill Pond. The park is located on the south side of Main Street, across from the pond. Park on either side of the road.

It is only a 75-foot walk to reach the rock, which can be readily spotted from the road.

Resources

Three Village Historical Society, *The Setaukets, Old Field, and Poquott*, Images of America Series (Charleston, SC: Arcadia Publishing, 2005). On page 107 is a photograph taken in 1927 of a number of women belonging to the Daughters of the Revolution standing in front of the rock.

"Patriot Rock: Historic Site, Setauket-East Setauket." https://foursquare.com/v/patriot-rock/4fa2c2c1e4b04db7bdfee77e/photos. This website contains several photographs of Patriot's Rock.

Raymond E. Spinzia, Judith A. Spinzia, and Kathryn E. Spinzia, *Long Island: A Guide to New York's Suffolk and Nassau Counties* (New York: Hippocrene Books, 1991), 49.

Howard Klein, *Three Village Guidebook: The Setaukets, Poquott, Old Field and Stony Brook*, 2nd ed., illus. Patricia Windrow (East Setauket, NY: Three Village Historical Society, 1986), 26.

Julia S. Smith, "Old Setauket," *Long Island Historical Society Quarterly* 2, no. 2 (April 1940): 48. Smith writes, "We find within the old town one monument which stood and still stands as it did, ages before the white man settled here or even the Indian social life began: the Rock, which was recently marked by Daughters of the Revolution."

"Patriot's Rock Historical Marker" https://www.hmdb.org/m.asp?m=114787.

Indian Rock 88

Type of Formation: Medium-Sized Boulder
WOW Factor: 3–4
Location: East Setauket (Suffolk County)
Tenth Edition, NYS Atlas & Gazetteer: p. 112, A5
Earlier Edition, NYS Atlas & Gazetteer: p. 26, A3–4
Indian Rock GPS Coordinates: 40°56.3717′N, 73°07.000′W
Accessibility: Roadside
Degree of Difficulty: Easy

Description

In *Three Village Guidebook: The Setaukets, Poquott, Old Field and Stony Brook*, Howard Klein writes, "This six foot by 15 foot boulder of fine gneiss, a metamorphic rock formed under great pressure and heat from other types of rock, contains the minerals feldspar, quartz, mica and others. It has been estimated to be about 200 million years old and was deposited here by the Wisconsin glacier."

History

Indian Rock's name is hardly unique or distinctive, much like the names Split Rock, Spook Rock, Spy Rock, or Balanced Rock, which tend to be generic and common in the rock community's vernacular. Perhaps a Native American presence is associated with the rock, which would account for its name.

Directions

From Stony Brook (junction of Routes 25A/North Country Road & 97/ Nicholls Road), head northeast on Route 25A for >1.3 miles. Turn left onto Main Street and proceed northwest for 0.6 mile until you reach Setauket Mill Pond. Just before crossing the pond, bear left, following Main Street as it now heads south. Go another 0.4 mile. The boulder is on the right side of the road, next to the third house after Lake Street. If you come to Watson Lane, on your left, then you have gone too far.

Resources

Howard Klein, *Three Village Guidebook: The Setaukets, Poquott, Old Field and Stony Brook*, 2nd ed., illus. Patricia Windrow (East Setauket, NY: Three Village Historical Society, 1986), 61.

David Weld Sanctuary Boulders 89

David Weld Sanctuary

Type of Formation: Large Boulder; Kettle Hole
WOW Factor: 4
Location: Nissequogue (Suffolk County)
Tenth Edition, NYS Atlas & Gazetteer: p. 112, B4
Earlier Edition, NYS Atlas & Gazetteer: p. 26, A3
Parking GPS Coordinates: 40°54.316′N, 73°12.517′W
Destination GPS Coordinates: *Beach Boulder*: 40°54.648′N, 73°12.643′W; *Kettles*: 40°54.302′N, 73°12.020′W (estimated)
Accessibility: *Beach Boulder*: 0.5–0.7-mile hike; *Kettles*: 1.0-mile hike
Degree of Difficulty: Moderate
Additional Information: *Self-Guided Tour brochure*: pbisotopes.ess.sunysb.edu/esp/Science_Walks/Weld/Weld.htm

---

Description

A number of large glacial erratics (probably in the 8–10-foot range) dot the landscape along the shore and in the woods.

A 60-foot-deep depression known as Kettle Hole was formed by a block of ice that melted in place at the end of the last glaciation.

History

The 125-acre sanctuary was donated to the Nature Conservancy by Mr. and Mrs. David Weld from 1969 to 1979. The Millers and the Woodys—neighbors of the Welds—also contributed by donating additional land.

The sanctuary borders the Long Island Sound and contains 1,800 feet of beachfront.

There are other notable water-filled kettles on Long Island that are listed by Margaret M. Voelbel in her book, *The Story of an Island: The Geology and Geography of Long Island*, these being: Lake Success (40°45.833'N, 73°42.497'W) near Great Neck; Lake Ronkonkoma (40°49.670'N, 73°07.353'W) in the town of Brookhaven; Artist Lake (40°53.043'N, 72°55.900'W) near Middle Island; Swan Lake (40°54.125'N, 72°47.593'W) near Manorville; Great Pond, aka Wildwood Lake (40°53.695'N, 72°40.469'W) near Riverhead; Water Mill Pond (40°54.877'N, 72°21.452'W) in the town of Southampton; Scuttle Hole (40°56.199'N, 72°49.850'W) near Bridgehampton; and Poxaboque (40°56.693'N, 72°17.117'W) in Bridgehampton.

KETTLE HOLE

The self-guided tour brochure explains in simple, clear language how a kettle hole comes into existence: "You can visualize how a kettle hole forms by imagining a block of ice put into an empty box. Fill the box with sand covering the ice. When the ice melts the water drains to the bottom of the box and a depression is left in the sand."

Directions

From Smithtown (junction of Routes 25A North/Main Street & 111/Hauppauge Road), drive north on Route 25A North for 250 feet. Bear left onto River Road and proceed north for 3.6 miles. When you come to Moriches Road, turn left and then immediately right onto Horse Race Lane. After driving north for 0.4 mile, bear left onto Short Beach Road and head west. In 0.1 mile, you will come to the parking area, on your right.

SHORELINE BOULDERS

Follow the white-diamond trail markers north for 0.5 mile to reach the beach and boulders along the shoreline.

LARGE INLAND BOULDER

Instead of proceeding north at the first junction, turn right and follow the trail east for >0.2 mile. When you come to the second junction, turn left (instead of right to reach the Kettles) and walk north until you come to a large boulder.

KETTLES

Follow the white-diamond markers north for >0.2 mile. At a junction, go right and head east for >0.2 mile. At another junction, bear right, and follow the trail, now taking you south, for 0.5 mile to reach the kettles.

Resources

"David Weld Sanctuary." https://www,nature.org/en-us/get-involved/how-to-help/places-we-protect/long-island-david-weld-sanctuary.

Margaret M. Voelbel, *The Story of an Island: The Geology and Geography of Long Island* (Port Washington, NY: Ira J. Friedman, 1965), 41.

Short Beach Boulder 90

Short Beach Park

Type of Formation: Large Boulder
WOW Factor: 7
Location: Nissequogue (Suffolk County)
Tenth Edition, NYS Atlas & Gazetteer: p. 112, B4
Earlier Edition, NYS Atlas & Gazetteer: p. 26, AB3
Parking GPS Coordinates: 40°54.382′N, 73°13.304′W
Short Beach Boulder GPS Coordinates: 40°54.515′N, 73°13.095′W
Accessibility: 0.3-mile walk
Degree of Difficulty: Easy
Additional Information: Short Beach Park, Short Beach Road, Nissequogue, New York 11780

Description

The Short Beach Boulder is roughly 15 feet high and 20 feet long, resting on the part of the beach that faces Long Island Sound.

History

The south side of Short Beach Park faces the Nissequogue River—an 8.3-mile-long river that rises from Smithtown. What's interesting about the Nissequogue River, like all the freshwater rivers on Long Island, is that it rises entirely from groundwater and not from a lake or pond.

Directions

Follow the directions given to the David Weld Sanctuary. Instead of turning right into the David Weld Sanctuary entrance, continue west on Short Beach Road for another 0.8 mile until you come to the parking area for Short Beach Park.

From the parking area, walk north for 0.1 mile and then bear right, following the shoreline east for 0.2 mile to reach the boulder.

We must confess that a considerable amount of time was spent initially looking for the Short Beach Boulder on Short Beach at Short Beach Island (located between Long Beach and Jones Beach State Park). Fortunately, we finally realized that this was the wrong Short Beach and reoriented ourselves to the correct spot, which is on the north side of Long Island—not the south.

Resources

"Short Beach Park, Short Beach Road, Nissequogue, New York." https://www.mapquest.com/us/new-york/short-beach-park-483406634. This site shows the location of the park relative to the ocean and Nissequogue River.

Port Jefferson Boulder 91

Type of Formation: Large Boulder
WOW Factor: 5
Location: Port Jefferson (Suffolk County)
Tenth Edition, NYS Atlas & Gazetteer: p. 113, A6
Earlier Edition, NYS Atlas & Gazetteer: p. 26, A4
Port Jefferson Boulder GPS Coordinates: 40°56.235′N, 73°03.503′W
Accessibility: Roadside
Degree of Difficulty: Easy
Additional Information: Village Internal Medicine Group, 710 Main Street, Port Jefferson, NY 11777

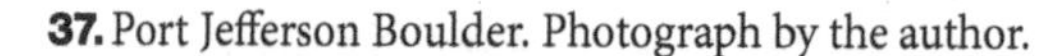
37. Port Jefferson Boulder. Photograph by the author.

Description

Danielle Mulch and Gilbert N. Hanson, in their article titled "Port Jefferson Geomorphology," call attention to a "truck-sized glacial boulder in Port Jefferson, N.Y. (a parking lot was built around it, given its size)."

In research done by Waldemar Patcholik and G. N. Hanson in 2001, called "Boulders on Stony Brook Campus May Reveal Geology of Long Island Sound Basement," three large boulders, partially sunk in the ground, were identified in the Port Jefferson valley, so there are certainly more boulders to be found if you wish to go off on your own and explore this area further.

History

What we found inspiring about the Port Jefferson Boulder is that a parking lot was built around it to accommodate the boulder instead of having the rock blown up to make room for the parking lot. High marks for civic-minded Port Jefferson.

Directions

From near the harbor in Port Jefferson (junction of Broadway & Main Street), head southeast on Main Street/Route 25A for 0.9 mile. Look for the boulder in a parking lot on the west side of Main Street between Reeves Road and High Street.

The rock lies next to the Village Internal Medicine Group.

Rocky Point Boulder 92

Type of Formation: Large Rock
WOW Factor: 9
Location: Rocky Point (Suffolk County)
Tenth Edition, NYS Atlas & Gazetteer: p. 113, A7
Earlier Edition, NYS Atlas & Gazetteer: p. 27, A5
Rocky Point Boulder Coordinates: 40°57.663′N, 72°56.327′W
Accessibility: Roadside

Description

The voluminous Rocky Point Boulder is a 3-story-high rock that is historically significant, for most likely the town of Rocky Point was named for it.

In 1905 the *Brooklyn Standard Union* newspaper described "a remarkable rock, a huge boulder, whose shadow sweeps over many rods of ground as the sun swings in its course, a source of astonishment to every visitor who sees it for the first time. Fifty feet long, forty feet thick and rising thirty-five feet above the ground."

According to a recent article by the editorial board of *Newsday*, the boulder may be the second largest glacial erratic on Long Island.

In "The Geology of Long Island," Jay T. Fox talks about a glacial boulder, 25 feet by 10 feet, in the vicinity of Hallock Landing. This second boulder is decidedly smaller than the Rocky Point Boulder, and therefore harder to find. So far, its location has eluded us.

History

The Rocky Point Boulder is also known to locals as Indian Rock, a name that arose when arrowheads were found in its vicinity.

The rock is on land originally owned by Noah Hallock. The Hallock farmhouse was built in 1721 and lived in by eight generations of descendants. The farmhouse is located nearby at 172 Hallock Landing Road. For the record, the Hallocks built one of their barns next to the enormous boulder.

Directions

From southeast of Sound Beach (junction of Routes 25A/North Country Road & 21/Rocky Point Yaphank Road & Rocky Point Road), drive north on Hallock Landing Road for 0.5 mile. At a traffic light, continue straight ahead on Hallock Landing Road for another 0.6 mile. Then bear left onto Sams Path (a road) and head west. Within 200 feet, you will see the rock, to your right, near the roadside, next to a resident's driveway.

As always in this kind of situation, do your looking from the roadside unless you have gotten permission from the landowner to approach the rock more closely.

Resources

"Giant Rock Is a Hard Case to Make in Suffolk County," editorial, *Newsday*, December 6, 2016, https://newsday.com/opinion/editorial/giant-rock-is-a-hard-case-to-make-in-suffolk-county-1.12707896.

"Historic Sites – Rocky Point Historical Society." http://rockypointhistoricalsociety.org/historic-sites.

"Suffolk Lawmaker Seeks to Preserve Island's Second-Largest Rock," *Newsday*, December 3, 2016, https://newsday.com/long-island/columnists/rick-brand/suffolk-lawmaker-seeks-to-preserve-island-s-second-largest-rock-1.12703056.

Indian Rock

93

Miller Place Beach

Type of Formation: Large Rock
WOW Factor: 6
Location: Miller Place (Suffolk County)
Tenth Edition, NYS Atlas & Gazetteer: p. 113, A7
Earlier Edition, NYS Atlas & Gazetteer: p. 27, A4–5
Parking GPS Coordinates: 40°57.713′N, 73°00.264′W
GPS Coordinates: *Indian Rock*: Unknown; *Rock #1*: 40°57.967′N, 73°00.154′W; *Rock #2*: 40°57.975′N, 72°59.650′W; *Millers Rock*: 40°57.996′N, 73°00.177′W
Accessibility: *Rock #1*: >0.3-mile walk; *Rock #2*: 0.8-mile walk
Degree of Difficulty: Moderately easy

38. Indian Rock. Antique postcard, public domain.

Description

Indian Rock is a sizeable boulder, perhaps 10 feet high or greater, resting against the foot of a sloping embankment. The postcard image accompanying this chapter should give you a good idea of what the rock looks like.

Millers Rock is a 20-foot island of rock that lies near the shoreline of Miller Place Beach.

History

Miller Place is located along the North Shore of Long Island and contains 2 miles of shoreline. In the late 1800s, Miller Place became a popular summer resort; later it proved attractive to seasonal residents. Its name comes from the Miller family, which included many of the area's initial settlers.

As far as we can tell, the Cordwood Landing County Park is the former Camp Francoise Barstow, a Girl Scout camp that operated until 1962.

Directions

The question for us was always how do you get to the beach without trespassing on private property? The following is what we've come up with. We believe you are allowed to access Cordwood Landing County Park without a permit as long as you are not intending to go fishing (which, we assume, you are not if your intention is just to look for the historic rock).

From northeast of Port Jefferson Station (junction of Routes 25A/Hallock Avenue & 347/Nesconset-Port Jefferson Highway), drive east on Route 25A for 2.4 miles. Turn left onto Miller Place Road/North Country Road and proceed north for 1.3 miles. Follow North Country Road as it veers left and continue west for another 0.4 mile. Bear right onto Landing Road and drive north for 0.2 mile. Finally, turn right onto Pringle Road (a dirt road) at the entrance to the Cordwood Landing County Park, indicated by a white-colored sign.

The shoreline is 0.3 linear mile from the parking area.

Rock #1 is located along the embankment down from the northeast corner of Cordwood Landing County Park. Rock #2 is roughly 0.5 mile farther east along the beach from the northeast corner of the park. If you're crossing beaches in front of homes, be sure to stay below the high-water mark.

We're not sure if either Rock #1 or Rock #2 qualifies for Indian Rock. You may need to do a little scouting, holding in your hand this book opened to the photograph of Indian Rock to assist in this venture.

Millers Rock lies close to the shoreline near the northeast end of the park.

Resources

Wikipedia, s.v. "Miller Place, New York," last modified December 2, 2022, http://en.wikipedia.org/wiki/Miller_Place,_New_York.

Wildwood Boulders 94

Wildwood State Park

Type of Formation: Medium-Sized Boulder
WOW Factor: 5–6
Location: Wading River (Suffolk County)
Tenth Edition, NYS Atlas & Gazetteer: p. 113, A8–9
Earlier Edition, NYS Atlas & Gazetteer: p. 27, A6
Parking GPS Coordinates: 40°57.865'N, 72°48.127'W
Destination GPS Coordinates: *Beach Boulder #1*: 40°57.992'N, 72°47.688'W; *Beach Boulder #2*: 40°57.990'N, 72°47.573'W; *Beach Boulder #3*: 40°57.996'N, 72°47.304'W; *Woods Boulder*: 40°57.813'N, 72°46.981'W
Fee: Modest fee charged
Hours & Days: Daily, sunrise to sunset
Accessibility: 0.3–0.6-mile walk
Degree of Difficulty: Easy
Additional Information: Wildwood State Park, 790 Hulse Landing Road, Wading River, NY 11792
Trail map: parks.ny.gov/documents/parks/WildwoodTrailMap.pdf

Description

A variety of boulders are encountered along a stretch of shoreline, both on land and partially in the water. According to Jessica L. McEachern and Daniel Davis in a paper entitled "Boulder Distribution at Wildwood State Park: Implications for Glacial Processes," "the dominant lithology of these boulders is granite and granite gneiss."

There are other boulders in the general area as well. In "The Geology of Long Island," Jay T. Fox writes about "2 large glacial boulders—one mile west of Wading River Station." That would place the boulders roughly near Lilco Road. Using Google Earth, we did see a good-sized boulder at

40°56.842′N, 72°51.967′W and, by good fortune, an old woods road/trail goes by it.

In the *New York Walk Book*, Raymond H. Torrey, Frank Place Jr., and Robert L. Dickinson write that "one of the largest of these erratics in the Harbor Hill Moraine lies near the Wading River marsh; it is a mass of reddish granite, originally twenty feet high and almost as wide and thick but has split into several fragments."

The Mountain Project website, mountainproject.com/area/116124416/wildwood-state-park, lists two areas in the park where large boulders can be found.

The Bluff Boulders are perched high on a bluff overlooking Long Island Sound: Delivery (12 feet high); Parachute Landing (15 feet high); Hillside Strangler (13 feet high); and Postal (13 feet high).

History

The Wildwood Boulders are part of the 600-acre Wildwood State Park, whose north side terminates at 50-foot-high bluffs overlooking Long Island Sound.

Directions

From east of East Shoreham (junction of Routes 25A & 46/William Floyd Parkway), drive east on Route 25A for 2.9 miles. As Route 25A veers right, continue east on Sound Avenue for 1.1 miles. Then turn left onto Route 54/Hulse Landing Road and head northwest for 0.9 mile. The park entrance is on your right. Drive east for 0.5 mile to the large parking area.

Follow a 0.1-mile-long path down to the beach. Bear right at the snack stand/restrooms and walk east for 0.6 mile to begin encountering the boulders. Some are slightly offshore, protruding out of the sea.

Resources

Raymond H. Torrey, Frank Place Jr., and Robert L. Dickinson, *New York Walk Book*, 3rd ed. (New York: American Geographical Society, 1951), 55.

"Sky and Stone: Long Island Bouldering – Wildwood State Park." https://skyandstone.blogspot.com/2008/11/long-island-bouldering-wildwood-state.html.

"Wildwood State Park Bouldering." https://www.mountainproject.com/area/116124416/wildwood-state-park.

Hulse Landing Beach Boulder 95

Hulse Landing Beach

Type of Formation: Large Boulder
WOW Factor: 6
Location: Wildwood (Suffolk County)
Tenth Edition, NYS Atlas & Gazetteer: p. 113, A9
Earlier Edition, NYS Atlas & Gazetteer: p. 27, A6
Possible **Parking GPS Coordinates:** 40°57.966′N, 72°48.677′W
Hulse Landing Beach Boulder GPS Coordinates: 40°58.065′N, 72°48.685′W
Accessibility: 200-foot-walk from end of road
Degree of Difficulty: Easy
Additional Information: A parking permit from the town of Riverhead is required from February 1 to December 31. It can be obtained at the Riverhead Parks and Recreation offices or online at riverheadrecreation.com/Course Activities.aspx?id=24&cat=18.

Description

The 15-foot-high boulder at Hulse Landing Beach is made of pegmatite, which is a coarse-grain igneous rock. The boulder lies directly on the shoreline, which ensures that it is constantly exposed to breaking waves.

History

The Hulse Landing Beach Boulder is found along the Roanoke Point Moraine—one of several places where glacial advancement stalled and boulders and rocks were deposited.

Directions

From east of East Shoreham (junction of Routes 25A & 46/William Floyd Parkway), drive east on Route 25A for 2.9 miles. As Route 25A veers right, continue east on Sound Avenue for 1.1 miles. Then turn left onto Route 54/Hulse Landing Road and head northwest for <1.3 miles.

The Hulse Landing Beach is located at the end of Hulse Landing Road.

Resources

"Long Island Glacial Erratic." https://epod.usra.edu/blog/2007/08/long-island-glacial-erratic.html.

Baiting Hollow Boulders 96

Type of Formation: Large Boulder
WOW Factor: 4–5
Location: Baiting Hollow (Suffolk County)
Tenth Edition, NYS Atlas & Gazetteer: p. 113, A10
Earlier Edition, NYS Atlas & Gazetteer: p. 27, A6–7
Parking GPS Coordinates: 40°58.625'N, 72°42.461'W
Destination GPS Coordinates: *First boulder:* 40°58.656'N, 72°42.455'W; *Second Boulder*: 40°58.687'N, 72°42.258'W; *Southwest Boulder*: 40°58.638'N, 72°42.526'W
Accessibility: >0.2-mile walk
Degree of Difficulty: Easy
Additional Information: A parking permit from the town of Riverhead is required from February 1 to December 31. It can be obtained at the Riverhead Parks and Recreation offices or online at riverheadrecreation.com/CourseActivities.aspx?id=24&cat=18.

Description

A number of medium-to-large-sized boulders are encountered along the beach in the Baiting Hollow area. Batting Hollow itself is tiny, only 3.2 square miles in size.

Directions

From east of East Shoreham (junction of Routes 25A & 46/William Floyd Parkway), drive east on Route 25A for 2.9 miles. As Route 25A veers right, continue straight ahead east on Sound Avenue for > 6.5 miles. Turn left onto Roanoke Avenue and head northwest for 0.8 mile to a parking area.

Walk north for 100 feet to reach the beach. You will see a medium-sized boulder immediately to your left, as well as several along the shoreline.

Follow the shoreline northeast, passing by a number of boulders and reaching a fairly large rock at 0.2 mile.

Large rocks can also be seen along the beach <0.1 mile southwest of the parking area.

SECOND OPTION

Three miles southwest of Roanoke Avenue near the end of Edwards Avenue [40°57.930'N, 72°46.179'W] are a number of decent-sized beach boulders at the GPS coordinates of 40°57.960'N, 72°46.027'W; 40°57.977'N, 72°45.883'W; and 40°57.983'N, 72°45.846'W. All look to be accessible from the beach. Parking, however, once again is by permit only. If you gain access, be sure to walk below the high-water mark if you are crossing in front of private homes.

Resources

"Baiting Hollow/Riverhead Area Bouldering." https://www.mountainproject.com/area/116124721/baiting-hollowriverhead-area.

Turtle Rock 97

Long Island Pine Barrens

Type of Formation: Large Boulder
WOW Factor: 6
Location: Ridge (Suffolk County)
Tenth Edition, NYS Atlas & Gazetteer: p. 113, B8
Earlier Edition, NYS Atlas & Gazetteer: p. 27, AB5–6
Parking GPS Coordinates: 40°52.503′N, 72°54.373′W
Turtle Rock GPS Coordinates: 40°52.548′N, 72°54.499′W
Accessibility: 0.1-mile hike
Degree of Difficulty: Easy

Description

Turtle Rock is a 10–12-foot-high, stand-alone boulder that is well known to partiers. Graffiti and broken glass indicate a long history of the rock being treated with disrespect.

History

Turtle Rock is located in the Long Island Pine Barrens, aka Long Island Central Pine Barrens, a publicly protected pine barren that covers more than 106,000 acres of land. It is considered to be Long Island's largest and last remaining natural area.

Directions

From the town of Ridge (junction of Route 25/Middle County Road & Route 46/William Floyd Parkway), drive southwest on Route 25/Middle County Road for 1.1 miles. Bear left onto Smith Road and head south for 1.1 miles. Then turn into a sandy pull-off on your right.

Follow a well-worn path west, bearing left when the trail splits. You will reach the boulder in 0.1 mile.

Resources

"Turtle Rock Bouldering." https://www.mountainproject.com/area/116124692/turtle-rock.

"Long Island Climbing." https://www.mountainproject.com/area/116124293/long-island.

Our Lady of the Island Boulder 98

Shrine of Our Lady of the Island

Type of Formation: Large Boulder
WOW Factor: 6–7
Location: Manorville (Suffolk County)
Tenth Edition, NYS Atlas & Gazetteer: p. 113, B9
Earlier Edition, NYS Atlas & Gazetteer: p. 27, AB6–7
Parking GPS Coordinates: 40°50.970′N, 72°45.420′W
Our Lady of the Island Boulder GPS Coordinates: 40°50.936′N, 72°45.526′W
Accessibility: 0.1-mile walk
Degree of Difficulty: Easy
Additional Information: Our Lady of the Island, 258 Eastport Manor Road, Manorville, NY 11949

Description

This large, 20-foot-high boulder, earlier known as the Rock Hill Boulder, has been repurposed to serve as the base of a statue of a woman holding a baby—a dedication to Mary, Queen of all Hearts.

In "A Study of Erratics on the Ronkonkoma Moraine in Eastport, Long Island," Diane Starbuck-Ribaudo, William Corbet, and Ann Marie Fishwick write that the boulder was first described by amateur geologist Colonel Bryson in 1895, who determined that the rock measured 50 feet by 20 feet. He called it the Rock Hill Boulder. Bryson estimated that the boulder must have been more than 125 feet by 20 feet before the rock was quarried.

Several smaller boulders are also mentioned in the Ribaudo-Corbet-Fishwick paper. Most are near the roadside, none being greater than 3 feet in height. You will see some of them as you drive or walk around the Our Lady of the Island grounds.

History

The Shrine of Our Lady of the Island was established by the Missionaries of the Company of Mary (the Montfort Fathers). In 1953, 70 acres of land in Eastport were donated by Crescenzo Vigliotta Sr. and Angelina Vigliotta for a shrine. In 1957, Mr. and Mrs. John Harrison donated the rock and surrounding acres overlooking Moriches Bay.

Directions

From I-495/Long Island Expressway, take Exit 70 for Manorville, Eastport & Route 111. Head southeast on Route 111/Eastport Manor Road/Captain Daniel Roe Highway for 2.0 miles. Turn right onto Eastport Manor Road where a green sign points the way to the shrine and follow it southeast for over 0.5 mile. Then bear right into the entrance boulevard to Our Lady of the Island and drive south for 0.5 mile to the parking area directly in front of the shrine's complex of buildings. If these parking spaces are occupied, there are many other spots to choose from along the drive in.

From the east side of the shrine, follow a wide path east for 0.1 mile to reach the boulder and statue and an open chapel.

Resources

J. Bryson, "Rock Hill, Long Island, N.Y.," *American Geologist* 16 (1895): 228–233.

Diane Starbuck-Ribaudo, William Corbet, and AnnMarie Fishwick, "A Study of Erratics on the Ronkonkoma Moraine in Eastport, Long Island" (unpublished manuscript, n.d.), https://dspace.sunyconnect.suny.edu/bitstream/handle/1951/47814/starbuck.pdf. On page 9 of the paper is a "map of Our Lady of the Island Shrine with locations of boulders studied." These lesser boulders appear to be about 3 feet in height.

Five Boulder Train and Kissing Rock

99

Shelter Island

Type of Formation: Large Boulder
WOW Factor: *Five Boulder Train*: 4–5; *Kissing Rock*: 5
Location: Shelter Island (Suffolk County)
Tenth Edition, NYS Atlas & Gazetteer: p. 114, B4
Earlier Edition, NYS Atlas & Gazetteer: p. 28, B3
General Parking GPS Coordinates: 41°04.289′N, 72°22.853′W
Destination GPS Coordinates: *Five Boulder Train*: 41°04.262′N, 72°22.854′W; *Kissing Rock*: 41°04.183′N, 72°23.068′W
Ferry Launch GPS Coordinates: *Tyndal Point*: 41°02.372′N, 72°18.923′W; *Greenport*: 41°06.010′N, 72°21.712′W
Ferry Landing GPS Coordinates: *South Ferry Hills*: 41°02.666′N, 72°19.032′W; *Shelter Island Heights*: 41°05.237′N, 72°21.494′W
Accessibility: Variable
Degree of Difficulty: Easy
Additional Information: Quinipet Camp and Retreat Center, 99 Shore Road, Shelter Island Heights, NY 11965; (631) 749-0430

Description

Five Boulder Train consists of an alignment of five boulders, each one containing a word of inspiration: "Love," "Faith," "Honesty," "Courage," and "Humility."

The word *quinipet*, from which the name of the Quinipet Camp and Retreat Center comes, is Latin for "five rocks."

The expression "boulder train" is used by geologists to describe a linear or fan-shaped distribution of boulders brought to their present location by glaciers. In *Rockachusetts: An Explorer's Guide to Amazing Boulders of Massachusetts*, Christy Butler and we were able to draw readers' attention to the Babson Boulders in Gloucester, Massachusetts, which are a perfect example of a boulder train. The one on Shelter Island is very much like the Babson Boulders, only on a much smaller scale.

Kissing Rock is a large, "5-million-pound chunk of granite" boulder, according to an article in the *New York Times* by John Rather, which is located on the beach at the end of Rocky Point Avenue, northeast of the Quinipet Camp and Retreat Center. "9/11 Rock" may be a new name for the boulder after it was redecorated following September 11, 2001.

History

The 25-acre Quinipet Camp and Retreat Center was founded in 1922. It is owned and operated by the New York Conference of the United Methodist Church.

According to Kent G. Lightfoot, Robert Kalin, and James Moore in *Prehistoric Hunter-Gatherers of Shelter Island, New York*, Shelter Island is "the largest of the glacially-created islands situated in the Peconic and Gardiners Bay of eastern Long Island." It formed around 18,000–23,000 years ago.

Per Phyllis Wallace, Shelter Island Historical Society archivist, Kissing Rock got its name in the 1940s and 1950s as a favorite trysting spot in the days before the area was developed and became more populated.

A ROCK STORY

In 2012, a potato-shaped, 60-ton boulder was shipped to Shelter Island. The mammoth boulder was delivered to 179 Ram Island Drive on Ram Island, whose owners used it as the end point of a 45-foot-long concrete bridge from the second floor of their house to the rock. It's not often that large boulders are used so practically and creatively.

Directions

There is no way to reach 8,000-acre Shelter Island without taking a ferry or arriving by boat. If you go by ferry, take either the North Ferry, which departs from the terminus of Route 114 at Greenport, or the South Ferry, which departs from the terminus of Route 114 near Tyndal Point.

Route 114/South Ferry Road, the main highway, bisects the island as it extends north/south.

The Quinipet Camp and Retreat Center is situated on the west side of the island at Jennings Point. The Five Boulder Train is located on the property of Quinipet Camp and Retreat.

Kissing Rock/9/11 Rock is located at the end of Rocky Point Avenue on the west side of the island.

If you walk north and then east from Kissing Rock, you will pass by a number of medium-sized boulders that lay strewn about along the beach on the property of the Quinipet Camp and Retreat Center.

As always, respect private property and be sure not to trespass without permission if land is posted.

Resources

"Shelter Island – Wikitravel." https://wikitravel.org/en/Shelter_Island. Kissing Rock is briefly mentioned.

Kent G. Lightfoot, Robert Kalin, and James Moore, *Prehistoric Hunter-Gatherers of Shelter Island, New York* (Berkeley, CA: Archaeological Research Facilities, 1987).

John Rather, "Shelter Rock Facing an Uncertain Future," *New York Times*, February 28, 1999, https://nytimes.com/1999/02/28/nyregion/shelter-rock-facing-an-uncertain-future.html.

East Marion Boulders 100

Southold Town Beach: Rocky Point

Type of Formation: Large Boulder
WOW Factor: 6
Location: East Marion (Suffolk County)
Tenth Edition, NYS Atlas & Gazetteer: p. 114, A4
Earlier Edition, NYS Atlas & Gazetteer: p. 28, AB3
Parking GPS Coordinates: 41°08.358'N, 72°21.135'W
East Marion Boulder GPS Coordinates: *Boulder along Rocky Point Road*: 41°08.301'N, 72°21.097'W; Heading west: *Water Rock*: 41°08.354'N, 72°21.287'W; *Second Boulder*: 41°08.351'N, 72°21.285'W; Heading east: 41°08.408'N, 72°20.956'W
Accessibility: <1.0-mile walk in totality
Degree of Difficulty: Moderately easy
Additional Information: A parking permit is required. Contact the Southold town clerk at Southold Town Clerk's Office, P.O. Box 1179, Southold, NY 11971.

Description

A number of large, 15-foot-high boulders can be seen along the beach and near the stairs at the end of Rocky Point Road.

In *Rockhounding New York: A Guide to the State's Best Rockhounding Sites*, Robert D. Beard writes, "The beach also has many large boulders of gneiss rock. It is continually reworked by the waves."

History

In the 1850s, East Marion was named for General Francis Marion, the "Swamp Fox" of the Revolutionary War. The town was earlier known as Oysterponds Upper Neck.

39. Bouldering East Marion Boulder. Photograph by Christian Prellwitz.

Directions

From north of Greenport (junction of Route 25 & Main Road), drive west on Route 25 for 1.4 miles. Turn left onto Rocky Point Road and head northwest for 1.1 miles. Take note of a medium-sized, roadside boulder on your right at 1.0 mile. When you come to the end of the road, park in the small area provided (permit required).

From the parking area, follow a short path north to the beach, taking note of a couple of large boulders at the end of the stairs.

West: From the stairs, turn left and walk west along the beach, coming to two large rocks in >0.1 mile, one along the shore and one out in shallow water.

East: From the stairs, turn right and walk east along the beach. One medium-sized boulder is reached after 200 feet, a second at 0.2 mile.

In his book *Rockhounding New York: A Guide to the State's Best Rockhounding Sites*, Robert D. Beard makes a worthwhile point when he states, "Keep in mind that while the beach below the high-water mark may be public, the surrounding property is private."

Resources

Frederick S. Lightfoot, Linda B. Martin, and Bette S. Weidman, *Suffolk County, Long Island in Early Photographs, 1867–1931* (New York: Dover Publications,

1984). On page 52 is a 1905 photograph of fishermen standing between several medium-sized boulders by the ocean.
Robert D. Beard, *Rockhounding New York: A Guide to the State's Best Rockhounding Sites* (Guilford, CT: Falcon Guides, 2014), 25. On page 24 is a photograph of Rocky Point beach and its large boulders.
"East Marion Boulders Climbing." https://www.mountainproject.com/area/109773979/east-marion-boulders.

Indian Memorial Rock Carvings 101

Type of Formation: Rock Carving; Medium-Sized Boulder
WOW Factor: 4
Location: Orient (Suffolk County)
Tenth Edition, NYS Atlas & Gazetteer: p. 114, A5
Earlier Edition, NYS Atlas & Gazetteer: p. 28, AB3
Parking GPS Coordinates: 41°08.921'N, 72°18.260'W
Destination GPS Coordinates: *Large Boulder*: 41°09.017'N, 72°18.143'W; *Indian Memorial Rock Carvings*: 41°09.355'N, 72°17.446'W
Accessibility: 0.9-mile trek
Degree of Difficulty: Moderate
Additional Information: A parking permit is required. Contact the Southold town clerk at Southold Town Clerk's Office, P.O. Box 1179, Southold, NY 11971.

Description

A series of nearly 25 carvings can be seen in a grouping of boulders along the Long Island Sound shoreline. Most of the etchings are of Native American faces.

History

The rock carvings were made, beginning in 1933, by Elliot A. Brooks, a local artist and fisherman, to commemorate Poquatuck and Montauk Native Americans. Most of the carvings were of Native Americans, but a few included other images. One is of a large polar bear. Altogether, a total of 25 images were carved in the rocks over a seven-year period. During his time, Brooks's outdoor workshop was called the Bear Rock Studio.

Nearly a hundred years of erosion have taken their toll on the rock images and probably in another hundred years they will be gone. However, photographs of the carvings can be seen in the Oysterponds Historical

Society Museum [41°08.269'N, 72°18.224'W], which is located on Village Lane in Orient.

From the earliest days on, Orient was a landing point for Native Americans traveling between Connecticut and Long Island.

Directions

As you approach the northeast tip of Long Island while driving east toward Orient along Route 25/Main Road, turn left onto Youngs Road, 0.9 mile past the parking area for Truman Beach (and >0.1 mile before you come to the Oysterponds Elementary School on your left). Drive north on Youngs Road for 0.4 mile until you reach the end of the road, which faces out toward Long Island Sound. In order to park, you must have a beach permit.

From the parking area, follow a 75-foot-long path down to the beach and walk east (to your right) along the shoreline for 0.9 mile until you come to the rock carvings. You will pass by a large boulder at <0.2 mile. Groupings of medium-to-large-sized boulders are passed further along the trek. Stay below the high tide mark.

The optimal time to visit the rock carvings, obviously, is during low tide.

Resources

Raymond E. Spinzia, Judith A. Spinzia, and Kathryn E. Spinzia, *Long Island: A Guide to New York's Suffolk and Nassau Counties* (New York: Hippocrene Books, 1991), 259.

"Bear Rock Studio – Orient, New York – Atlas Obscura." https:www.atlasobscura.com/places/bear-rock-studio.

"Elliot A. Brook's Carvings." https://www.jeremynative.com/onthissite/listing/elliot-a-brooks-carvings.

"Orient's Treasure on the Edge of the Sea." https://www.peconicbathtub.com/orients-treasure-on-the-edge-of-the-sea.

Jacobs Point Boulder 102

Type of Formation: Large Boulder
WOW Factor: 8
Location: Northville (Suffolk County)
Tenth Edition, NYS Atlas & Gazetteer: p. 114, C1
Earlier Edition, NYS Atlas & Gazetteer: p. 27, A7
Iron Pier Beach Parking GPS Coordinates: 40°59.243′N, 72°36.949′W
Jacobs Point Boulder GPS Coordinates: 40°58.801′N, 72°39.612′W
Accessibility: *By water*: 2.3 mile-trek
Degree of Difficulty: Easy by boat
Additional Information: A parking permit from the town of Riverhead is required from February 1 to December 31. It can be obtained at the Riverhead Parks and Recreation offices or online at riverheadrecreation.com/Course Activities.aspx?id=24&cat=18.

Description

In "The Geology of Long Island," Jay T. Fox describes the giant boulder at Jacobs Point as "38 feet by 20 feet," literally the size of a large garage.

Directions

From west of Aquebogue (junction of Routes 105/Cross River Drive & 25/Main Road), drive northwest on Route 105/Cross River Drive for 2.1 miles. Turn right onto Sound Avenue and proceed east for 0.6 mile. Turn left onto Pennys Road and go northwest for 0.8 mile. At this point, take note that Pennys Road continues straight ahead, but leads to private residences that are very close to where Jacobs Point Boulder is.

Turn right onto Sound Shore Road and drive east for 2.3 miles. When you come to Pier Road, bear left, and then immediately right into the parking area for Iron Pier Beach.

Launch your watercraft from the boat launch and head west for 2.3 miles. You will see the Jacobs Point Boulder, just up from the beach, 60 feet in front of a private residence. That's probably as close as you can get.

Resources

"Jacobs Point – Natural Atlas." https://naturalatlas.com/points/jacobs-1865909.

Hither Hills Split Rock and Lost Boulder

103

Hither Hills State Park

Type of Formation: Split Rock; Large Boulder
WOW Factor: 5
Location: Montauk (Suffolk County)
Tenth Edition, NYS Atlas & Gazetteer: p. 115, BC9
Earlier Edition, NYS Atlas & Gazetteer: p. 29, BC5–6
Parking GPS Coordinates: 41°00.837'N, 72°00.409'W
Split Rock & Lost Boulder GPS Coordinates: Not determined
Hours: Sunrise to sunset
Accessibility: *Split Rock*: >0.8-mile hike; *Lost Boulder*: Not determined, but probably >2.0 mile-hike
Degree of Difficulty: Moderate
Additional Information: *Hither Hills State Park & Hither Woods Park map*: parks.ny.gov/documents/parks/HitherHillsTrailmap.pdf; Parking/hiking permits are free but required. They last for three years and can be obtained from the New York State Department of Environmental Conservation.

Description

Split Rock is a large boulder that has split diagonally in half, its two sections relatively close together. A third, much smaller section of the boulder lies on the ground, broken off from one of the halves. The trail goes directly past the boulder.

Lost Boulder is a 12-foot-high rock. Its name suggests that it is not the easiest boulder to find.

History

The 1,755-acre Hither Hills State Park was created in 1924 by the Long Island Park Commission to ensure that the property would not be lost to future land developers.

Hither Hills was named by early settlers who used the expression "come hither" as an invitation for all to enjoy the land.

Directions

From Amagansett (junction of Routes 27/Montauk Highway & 74/Abrahams Landing Road), drive east on Route 27/Montauk Highway for >6.0 miles. When you come to Old Military Road, on your right), continue northeast on Route 27 for another 1.0 mile and turn left into a substantial-sized parking area for the Hither Hills West Overlook.

The kiosk is located on the east side of the parking area.

SPLIT ROCK

Follow the Parkway Trail northeast as it crosses over the black-marked Serpents Back Path (SB) and Power Line Road (PLR). At 0.3 mile, turn left onto the yellow-marked Split Rock Road and head northwest for another 0.5 mile to reach Split Rock.

LOST BOULDER

Although the Lost Boulder Trail is listed on the park map, it is unlikely to be noticed by most hikers since only the initials LB (LB for "Lost Boulder") are provided. Furthermore, the boulder's location is not indicated on the map. What we can provide, then, are simply general directions: From Split Rock, continue northeast on the yellow-marked Split Rock Trail. At a junction, turn left onto the red-marked Ram Level Road (RL) and head north for 0.5 mile. At a junction, turn right onto the blue-marked Foggy Hollow Road (FHR) and proceed northeast. When you come to the white-marked Paumanok Path (PP), turn right and proceed east. Somewhere along this path, you will either pass by the Lost Boulder or reach a short spur path that diverts to it. We would imagine reaching the Lost Boulder is a hike of at least 2.0 miles.

Resources

C. R. Roseberry, *From Niagara to Montauk: The Scenic Pleasures of New York State* (Albany, NY: State University of New York Press, 1982), 309–311.

"Hither Hills State Park." https://parks.ny.gov/parks/122/details.aspx.

"Hither Hills and Hither Hills Loop." htpps://www.nynjtc.org/hike/hither-hills-and-hither-woods-loop#dialog-hike-description.West Side of Hudson River

West Side of Hudson River

Orange County

Orange County is named after King William III of England, who also held the titles of Stadtholder of Holland and Prince of Orange. The county encompasses 839 square miles, virtually all of it land.

It is the only county in New York State that borders both the Hudson River and the Delaware River. Its northeastern corner is part of the 700,000-acre Catskill Park.

Orange County is considered part of the Poughkeepsie–Newburgh–Middletown metropolitan statistical area.

Downing Park Rock 104

Downing Park

Type of Formation: Large Rock
WOW Factor: 4
Location: Newburgh (Orange County)
Tenth Edition, NYS Atlas & Gazetteer: p. 103, E5
Earlier Edition, NYS Atlas & Gazetteer: p. 36, D3–4
Parking GPS Coordinates: 41°30.286′N, 74°01.220′W (Third Street)
Downing Park Rock Coordinates: 41°30.354′N, 74°01.196′W
Hours: Daily, dawn to dusk
Accessibility: 0.1-mile walk
Degree of Difficulty: Easy

Description

The Downing Park Rock is a massive mound of bedrock 8 feet high and 20 feet long resting on a high point of land. A second, smaller, 4-foot-high and 10-foot-long mound of rock can be seen to the north, only 150 feet away.

There are rocky outcrops throughout the park but all are substantially smaller in size.

History

Downing Park opened in 1897 and encompasses 35 acres of rolling hills, with streams, a pond (informally known as "The Polly") with a fountain, and well-groomed paths. It is very much in keeping with the design of New York City's Central Park.

The park is named for horticulturist Andrew Jackson Downing, a native of Newburgh, and a pioneer of the public park movement.

Directions

From I-84 near the west end of the Newburgh-Beacon Bridge, take Exit 39 for Route 9W South & Highland. Drive south on Route 9W for 1.0 mile. You will see the park to your left. Turn left onto Third Street and park.

From Third Street, walk north, passing by the west side of the Downing Park Pond. In 0.1 mile, you will reach the large rock, poised on a high point of land.

This is a lovely park to stroll through, containing interesting features along its walkways.

Resources

"About – Downing Park." https:///www.downingparknewburgh.org/about.

"Downing Park." https://www.downingparknewburgh.org.

Kevin Barrett, *Newburgh*, Images of America Series (Charleston, SC: Arcadia Publishing, 2007). A collection of early twentieth-century photographs taken in the park are shown on pages 71–78.

Devil's Dance Hall (Historic) 105

Danskammer Point

Type of Formation: Historic Rock Protrusion
WOW Factor: 3
Location: Danskammer Point, Roseton (Orange County)
Tenth Edition, NYS Atlas & Gazetteer: p. 103, E7
Earlier Edition, NYS Atlas & Gazetteer: p. 36, D4
Danskammer Point GPS Coordinates: 41°34.424'N, 73°57.805'W
River Access Parking GPS Coordinates: *White's Hudson River Marina* (Dutchess County): 41°35.171'N, 73°57.025'W
Accessibility: 1.0-mile trek by water

Description

The Devil's Dance Hall, aka Danskammer Point, is a long, flat surface of bedrock that juts out into the Hudson River. In *Handy Guide to the Hudson River and Catskill Mountains*, Ernest Ingersoll describes the point as "a rocky headland with wall-like fronts of white rock." It was accidentally destroyed or greatly altered when the *Cornell*, a large steamer, rammed into it one misty morning in 1890.

History

Danskammer Point was named by Henry Hudson's crew in 1609 as their ship, the *Half Moon*, sailed up the Hudson River. Much to their astonishment, they observed in the early evening a cluster of Native Americans, painted in red, whooping and hollering, and dancing wildly around a flickering fire. It was like something out of Dante's *Inferno*. The word that Hudson's crew came up with, *danskammer*, is Dutch for "dance hall" or "dance chamber."

A similar account of Native Americans participating in a wild ceremony at Danskammer Point was told by Lieutenant Cowenhoven in 1663.

Captain Stanley Wilcox and H. W. Van Loan, in *The Hudson from Troy to the Battery*, state that there is wide belief that Captain Kidd buried pirate treasure on the point. If so, it has never been found despite the efforts of many who have poked around, only to leave the site empty-handed.

Ernest Ingersoll asserts that "this point was [once] the boundary-line between the jurisdictions of New Amsterdam and Fort Orange (Albany)."

In the 1820s, Edward Armstrong purchased Danskammer Point and by 1834 had built a house that overlooked the point. He named it Danskammer. Andrew Jackson Downing, the famous Newburgh architect, was hired to landscape the grounds. The house lasted until 1932, when it was demolished.

Meanwhile, Danskammer Point had its own issues. After it was severely damaged by a ship that accidentally struck it during the late nineteenth century, a 31-foot-high lighthouse was built on the rock fragments in 1885. In 1914, the lighthouse was hit by lightning and the lighthouse keeper, James H. Wiest, temporarily paralyzed from the strike. It almost did seem like Danskammer Point was cursed.

The Danskammer Point Lighthouse is gone, decommissioned in the 1920s, and most of the site is now occupied by the Danskammer Energy Generating Station.

Directions

LAND ACCESS

Land access appears unlikely due to the presence of the Danskammer Energy Generating Station, which now occupies much of the site away from the river.

WATER ACCESS

For those who wish to see what is left of Danskammer Point, launch your watercraft from White's Hudson River Marina on 15 Point Street in New Hamburg. *To get there*: From Route 9D, take New Hamburg Road northwest for 1.3 miles. Turn left onto Bridge Street, crossing over railroad tracks. At the end of Bridge Street, turn right onto Main Street and then immediately left onto Point Street, which takes you to the marina in 0.1 mile.

Now traveling by boat, head southwest down the Hudson River for 1.0 mile to reach Danskammer Point. Bear in mind that direct landing on the point may be prohibited or, minimally, restricted.

Resources

"History: Danskammer Point Light." https://www.cornwall-on-hudson.com/article.cfm?page=718.

Arthur G. Adams, *The Hudson River Guidebook* (New York: Fordham University Press, 1996), 203.

Lewis Beach, *Cornwall* (Newburgh, NY: E. M. Ruttenber & Sons, 1873), 8–9. Beach tells about the story at Danskammer Point and the frenzied dancing, and mentions how the boat captain came to exclaim "De Duyfel's Dans-Kammer!," meaning "The Devil's Dance Chamber."

Stanley Wilcox and H. W. Van Loan, *The Hudson from Troy to the Battery* (Philmont, NY: Riverview Publishing, 2011), 71–72.

Ernest Ingersoll, *Handy Guide to the Hudson River and Catskill Mountains* (Astoria, NY: J. C. & A. L. Fawcett, 1989; reprint of 1910 book), 128.

William F. Gekle, *The Lower Reaches of the Hudson River* (Poughkeepsie, NY: Wyvern House, 1982), 53.

Cornelia F. Bedell, *Now and Then and Long Ago in Rockland County* (Rockland County, NY: Privately printed, 1941), 41–42.

Lifting Rocks and Hawk's Nest Rock 106

Type of Formation: Perched Boulder
WOW Factor: 5
Location: Sparrow Bush (Orange County)
Tenth Edition, NYS Atlas & Gazetteer: p. 106, B3
Earlier Edition, NYS Atlas & Gazetteer: p. 31, AB4–5
Parking GPS Coordinates: 41°25.234′N, 74°44.021′W
Lifting Rocks and Hawk's Nest Rock GPS Coordinates: 41°25.257′N, 74°43.971′W (guesstimate)
Accessibility: Unknown
Degree of Difficulty: Difficult

Description

In *Hudson Valley Trails and Tales*, Patricia Edwards Clyne writes about "Lifting Rocks, a spectacular formation high above Hawk's Nest Drive (Route 97), a few miles west of Port Jervis."

In *Underground Empire: Wonders and Tales of New York Caves*, Clay Perry refers to the rocks as "a curious cavity formed by a huge flat slab supported on legs of stone," and writes that the "formation consists of a huge slab of rock held up by three stone legs, forming a shelf or slit into which a man might crawl sideways."

According to Michael J. Worden on his website michaeljworden.com/stones.html, three large, perched rocks form a triangular pattern, with each boulder being of considerable size. He considers the boulders to be megalithic in origin.

40. Hawk's Nest Rocks. Antique postcard, public domain.

History

Hawk's Nest, a rocky outcropping, probably got its name from birds of prey that nested in the area.

In her book, *Hudson Valley Trails and Tales*, Patricia Edwards Clyne tells the tale of Tom Quick, an eighteenth-century Indian fighter, who used the Lifting Rocks high ground as a lookout, and from it killed a party of Delaware warriors advancing from below. Clay Perry also narrates a similar story in *Underground Empire: Wonders and Tales of New York Caves.*

Directions

From Sparrow Bush (junction of Routes 97 & 42), drive northwest on Route 97 for 1.3 miles and turn left into a tiny parking area for the Hawk's Nest Lookout. Look for a state historic marker on top of the stone wall.

This is as far as we can take you. From here, you are on your own. We're assuming that the Lifting Rocks are uphill rather than downhill from the pull-off. This means, then, that to reach the top of the sloping hill will require a near vertical ascent of 175 feet, with the first 10 feet or so straight up. We don't believe the ascent is doable for anyone other than a rock climber.

An alternative approach to this upper reach would be to bushwhack through dense woods for 0.5 mile from Route 42 (assuming that the land isn't posted), but this is something that we would advise against doing unless you are very competent with a compass or GPS tracking unit.

There is a sizeable area of exposed bedrock [41°25.124'N, 74°43.946'W] 0.1 mile southeast (at the top of the hill) and this might bear a look.

On the other hand, the slight possibility exists that Lifting Rocks is directly below the lookout, where a grouping of rocks is visible on Google Earth. Once again, a little finesse may be required to get down to this lower level to determine exactly what is there.

Resources

Patricia Edwards Clyne, *Hudson Valley Trails and Tales* (Woodstock, NY: Overlook Press, 1990), 171.

Clay Perry, *Underground Empire: Wonders and Tales of New York Caves* (New York: Stephen Daye Press, 1948), 127–128.

"Mysterious Stone Sites in the Hudson Valley." http://www.michaeljworden.com/stones.html.

Megaliths and Caves

107

Schunemunk State Park: Schunemunk Mountain

Type of Formation: Block; Crevice
WOW Factor: 6
Location: Mountainville (Orange County)
Tenth Edition, NYS Atlas & Gazetteer: p. 107, AB10
Earlier Edition, NYS Atlas & Gazetteer: p. 32, A3
Parking GPS Coordinates: 41°24.457'N, 74°04.911'W
Megaliths & Caves GPS Coordinates: Not determined
Accessibility: >3.0-mile hike; 1,500-foot ascent
Degree of Difficulty: Moderately difficult
Schunemunk State Park Trail Map: parks.ny.gov/documents/parks/SchunnemunckTrailMap.pdf

---

Description

The Megaliths consist of large blocks of bedrock that are separated by deep crevices.

According to Peggy Turco, in *Walks and Rambles in the Western Hudson Valley*, "The Megaliths are a group of impressive conglomerate knobs that pulled away, it seems, from the main body of bedrock." Caves, or fissures, have formed in the spaces created by the separation.

A megalith, by definition, is a stone of great size. The word is typically used to describe ancient constructions, such as dolmens, which the Schunemunk Megaliths are clearly not.

History

The 1,664-foot high Schunemunk Mountain is the highest mountain in Orange County and also dissimilar geologically from its mountainous

41. Megaliths. Photograph by Dan Balogh.

cousins in the Hudson Highlands. Its name is Lenape for "excellent fireplace."

The mountain created unwanted notoriety in 2002 when a boulder being grasped by the lead hiker of a small group broke loose, killing him and then, rebounding, seriously injuring two others below.

In writing this book, we always assume that boulders are harmless and pleasant entities that nature has created, but one always needs to be mindful of their mass and destructive power when set into motion. In other words, be careful when you are hiking amid boulders or loose rock, especially if the terrain is becoming vertical.

Directions

From Vails Gate (junction of Routes 32, 300/Temple Hill Road, & 94), drive southeast on Route 32 for 4.2 miles. Turn acutely right onto Route 79/Pleasant Hill Road. Go 0.1 mile north and then turn left onto Taylor Road, which takes you immediately across Woodbury Creek. Stay straight (as Starr Road goes off to your left) and then within 0.1 mile cross over I-87/NYS Thruway. After 0.1 mile from the Thruway crossing, you will come to the parking area for Schunemunk Mountain located on both sides of the road. A kiosk is located on the northeast side of the parking area.

From the parking area, follow the white-blazed Sweet Clover Trail southwest. Soon you will cross railroad tracks and then begin climbing steadily uphill. At a junction, turn left and follow the yellow/teal-blazed Jessup Trail south. Continue past the junction with Dark Hollow Trail, which comes in on your left. Eventually, you will come to two large cairns where the word MEGALITH is painted on the bedrock in big, white letters. A large, white-colored arrow points the way to your right. Follow the spur trail to reach the Megaliths.

Resources

Peggy Turco, *Walks and Rambles in the Western Hudson Valley* (Woodstock, VT: Backcountry Publications, 1996), 50.

"Schunemunk Mountain." https://hikethehudsonvalley.com/hikes/schunemunk-mountain.

"Schunemunk Mountain Loop from Otterkill Road." https://www.nynjtc.org/hike/schunemunk-mountain-dark-hollow-jessup-trails.

Peter Kick, Barbara McMartin, and James M. Long, *50 Hikes in the Hudson Valley: From the Catskills to the Taconics, and from the Ramapos to the Helderbergs*, 2nd ed. (Woodstock, VT: Backcountry Publications, 2000), 69–72.

Stella Green and H. Neil Zimmerman, *50 Hikes in the Lower Hudson Valley* (Woodstock, VT: Backcountry Guides, 2002), 175.

Crow's Nest Boulder and Captain Kidd's Cave

108

Storm King State Park: Crow's Nest

Type of Formation: Large Boulder; Cave
WOW Factor: Not determined
Location: West Point (Orange County)
Tenth Edition, NYS Atlas & Gazetteer: p. 108, B2
Earlier Edition, NYS Atlas & Gazetteer: p. 32, AB3–4
Parking GPS Coordinates: 41°24.895'N, 73°59.623'W
Destination GPS Coordinates: Not determined
Accessibility: *Large Boulder*: 1.0-mile hike; *Captain Kidd's Cave*: Unknown

---

Description

In *Walks and Rambles in the Western Hudson Valley*, Peggy Turco mentions encountering "a large erratic" just before a grassy knoll is reached that provides the first views of the Hudson Valley.

A boulder is also encountered at the summit of Crow's Nest.

In Patricia Edwards Clyne's *Hudson Valley Trails and Tales*, a Captain Kidd's Cave located on Crow's Nest is mentioned, but no specific directions to it are provided. Clyne writes, "Oddly enough, a treasure of sorts was found in the Crow's Nest Cave, but it was not Captain Kidd's. In the fall of 1870, a group of explorers found several old coins, none dated earlier than 1782"—a realization that since these coins were minted nearly a century after Kidd's death, they couldn't possibly be part of his treasure stash.

Clyne's story is probably based on an account in Lewis Beach's 1873 book, *Cornwall*: "Quite a number of feet above the flow of the Hudson, in the rugged and precipitous breast-bone of 'Old Cro'-nest,' was discovered,

42. Breakneck Mountain from Crow's Nest. Photograph by Dan Balogh.

in the fall of 1872, a huge cavern." Beach goes on to narrate the story of how the coins were discovered and speculates on who might have left them.

Of particular interest is the manner in which James Fenimore Cooper adapted this eighteenth-century tale to his story about Enoch Crosby in his 1821 novel, *The Spy*.

In the *Hudson Highlands*, William Thompson Howell talks about several shelter caves, one of which may, in fact, be Captain Kidd's Cave. Bat Cave was found when Howell and his hiking companions left Sherwood's Rock and did an arduous climb up Crow's Nest: "We wiggled into the small mouth of the cave, and by the light of half a dozen candles explored the cavity." In another section, Lowell writes about "The Grotto on Cro' Nest": "The Grotto . . . is a long shallow cavern in the granite uplift of rocks . . . and beneath whose shelter quite a good many persons might take refuge from a storm."

There is also a formation on a high bluff above the western bank of the Hudson River that is known as Kidd's Plug Cliff. Most likely, this bluff is the 1,350-foot east face of Crow's Nest Mountain. In *The Hudson: From the Wilderness to the Sea*, Benson Lossing writes, "High up on the smooth face of the rock, is a mass slightly projecting, estimated to be twelve feet in diameter, and by form and position, suggesting, even to the dullest imagination, the idea of an enormous plug stopping an orifice." According to Arthur G. Adams, in *The Hudson River Guidebook*, the plug was "destroyed either by quarrying or construction of the Old Storm King Highway—or possibly by artillery practice."

History

According to E. M. Ruttenber and L. H. Clark, in *History of Orange County, New York*, Cro'-Nest's "modern name preserves in substance its Algonquin name, which, in ancient records, is written *Navesing*, signifying a 'resort for birds.'"

We would be remiss if we didn't mention that the mountain was used as target practice for testing Parrot guns manufactured across the river at the West Point Foundry in Cold Spring.

Directions

From west of West Point (junctions of Routes 9W, 293, & 218) drive north on Route 9W for >2.5 miles and turn into a pull-off on your right just before a yellow "icy pavement zone" sign. A kiosk at the back of the parking area provides helpful information about the area.

Follow the white-marked Bobcat Trail north for 0.5 mile. When you come to a junction, bear right, heading north and now following the blue-colored Howell Trail, named for William Thompson Howell, a late nineteenth-century writer and photographer.

Mention is made that a large glacial erratic awaits on the summit. Along the climb, you will pass by the boulder mentioned in *Walks and Rambles in the Western Hudson Valley*.

We have no idea where Captain Kidd's Cave is located on the mountain. We do know that access to part of the mountain is restricted because of a substantial number of nineteenth-century unexploded ordinances that still litter the landscape.

Resources

William Thompson Howell, *The Hudson Highlands*, 2 vols. (New York: Walking News, 1982).

Peggy Turco, *Walks and Rambles in the Western Hudson Valley* (Woodstock, VT: Backcountry Publications, 1996), 24.

Patricia Edwards Clyne, *Hudson Valley Trails and Tales* (Woodstock, NY: Overlook Press, 1990), 203, 250.
New York-New Jersey Trail Conference, *Day Walker: 32 Hikes in the New York Metropolitan Area*, 2nd ed. (Mahwah, NJ: New York-New Jersey Trail Conference, 2002), 224–229.
Benson Lossing, *The Hudson: From the Wilderness to the Sea* (Sommersworth, NY: New Hampshire Publishing Company, 1972; facsimile of the 1866 edition), 217–218. On page 217 is a line drawing of Kidd's Plug Cliff.
E. M. Ruttenber and L. H. Clark, *History of Orange County, New York*, vol. 1 (Interlaken, NY: Heart Lake Publishing, 1980), 33–34.
Ernest Ingersoll, *Handy Guide to the Hudson River and Catskill Mountains* (Astoria, NY: J. C. & A. L. Fawcett, 1989; reprint of 1910 book), 108–109.

Walt Whitman Boulder 109

Bear Mountain State Park

Type of Formation: Large Boulder
WOW Factor: 4
Location: Bear Mountain (Rockland County)
Tenth Edition, NYS Atlas & Gazetteer: p. 108, C2
Earlier Edition, NYS Atlas & Gazetteer: p. 32, B3–4
Parking GPS Coordinates: 41°18.736′N, 73°59.327′W
Walt Whitman Boulder GPS Coordinates: 41°19.009′N, 73°59.318′W
Fee: Modest admission charged
Hours & Days: Consult website for specific details
Accessibility: >0.3-mile walk
Degree of Difficulty: Easy
Additional Information: Bear Mountain State Park, Palisades Parkway/ Route 9W, North Bear Mountain, NY 10911
Trail Map: parks.ny.gov/documents/parks/BearMountainTrailMap.pdf
Northern Harriman Bear Mountain Trails, Trail Map 4

Description

The Walt Whitman Boulder, supporting a 9-foot-high statue of Walt Whitman, is roughly 6 feet high and 12 feet long—a respectable-sized rock.

History

The statue of Walt Whitman was designed by Jo Davidson and first exhibited at the 1939 New York World's Fair before arriving at its present location in 1940. The statue was commissioned by the Harriman family to honor their mother, Mary Williamson Harriman who, thirty years earlier, had donated land and money to establish the Bear Mountain-Harriman section of the Palisades Interstate Park.

43. Walt Whitman Boulder. Antique postcard, public domain.

A stanza from Whitman's "Song of the Open Road" is etched into one side of the boulder.

Interestingly, there is also a Walt Whitman Boulder on the summit of Jayne's Hill, located in the West Hills County Park on Long Island [40°48.918′N, 73°25.517′W]. The boulder is considerably smaller (probably 2–3 feet in height) and has been vandalized. A plaque on the West Hills County Park boulder is inscribed with another one of Whitman's poems, "Paumanok," from *Leaves of Grass.*

The Bear Mountain Bridge, spanning the Hudson River near the park, was opened in 1924. At the time, it was the largest single span bridge in the world.

Directions

From the Bear Mountain Bridge traffic circle at the west end of the Bear Mountain Bridge (junction of Routes 202, 9W, & 6), drive south on Route 202/9W for 0.4 mile and turn right at the first traffic light onto Seven Lakes Drive. Then bear right into the second driveway to reach the main parking lot.

From the northeast end of the parking area, walk north for <0.3 mile and then turn east, going down a staircase and following a tunnel under Route 9W that takes you to the east side of the road. From here, it is a walk of several hundred feet to the statue and boulder as you follow signs for the Bear Mountain Zoo and Trailside Museum.

Resources

Bill Bailey, *New York State Parks: A Guide to New York State Parks* (Saginaw, MI: Glovebox Guidebooks of America, 1997), 255.

"Walk Whitman Historical Marker." https://www.hmdb.org/m.asp?m=47774.

New York-New Jersey Trail Conference, *Guide to the Appalachian Trail in New York and New Jersey*. 9th ed. (Harpers Ferry, WV: Appalachian Trail Conference, 1983), 71.

Leonard M. Adkins and the Appalachian Trail Conservancy, *Along the Appalachian Trail: New Jersey, New York, and Connecticut*, Images of America Series (Charleston, SC: Arcadia Publishing, 2014), 75. A photograph of the Walt Whitman statue is shown but, unfortunately, very little of the underlying boulder is visible.

Turtle Rock 110

Warwick County Park

Type of Formation: Medium-Sized Rock
WOW Factor: 4
Location: Warwick (Orange County)
Tenth Edition, NYS Atlas & Gazetteer: p. 107, D7–8
Earlier Edition, NYS Atlas & Gazetteer: p. 32, BC1
Parking GPS Coordinates: 41°14.410′N, 74°19.926′W
Turtle Rock GPS Coordinates: 41°14.351′N, 74°19.708′W
Hours: Daily, 9:00 a.m. to 9:00 p.m.
Accessibility: 0.5-mile hike
Degree of Difficulty: Moderate
Additional Information: Warwick County Park, 25 County Park Lane, Route 17A, Warwick, NY

Description

This unique, medium-sized rock has achieved some notoriety due to its turtle-like shape. It is not a big rock, but its defining quality is that it is distinctive.

History

Some folks contend that the protrusion on the left side of Turtle Rock, which looks like the head of a turtle, was carved out by earlier visitors and is not naturally formed. Steve Schimmrich, the Hudson River geologist, however, is of a different opinion and contends that the formation is entirely natural. In this matter, we stand with the Hudson Valley geologist.

Warwick County Park encompasses 48 acres.

44. Turtle Rock. Photograph by Steven Schimmrich.

Directions

From south of Warwick (junction of Routes 17A East/Galloway Road & 94/Oakland Avenue), drive east on Route 17A East for <1.3 miles and turn right onto County Park Lane where a green-colored sign invites you to Warwick County Park. Proceed southeast for 0.5 mile. As County Park Lane veers left, continue straight ahead (south) for another 0.2 mile to reach the parking area.

From the parking area, follow a road south for 0.2 mile along the east side of the woods. Then turn left onto a trail that goes uphill into the woods, heading northeast. After negotiating a hairpin turn, you will come to the top of the hill, where Turtle Rock can be seen to your left, slightly off the path. By then, you will have hiked 0.5 mile.

Dutchess Quarry Caves

111

Type of Formation: Shelter Cave
WOW Factor: 5
Location: Finnegans Corners (Orange County)
Tenth Edition, NYS Atlas & Gazetteer: p. 107, B7
Earlier Edition, NYS Atlas & Gazetteer: p. 32, AB1
Mt. Lookout GPS Coordinates: 41°21.578′N, 74°21.664′W
Quarry Caves GPS Coordinates: 41°21.585′N, 74°21.817′W (per Funk and Steadman)
Accessibility: Unknown

Description

The Dutchess Quarry Cave Sites consist of a series of rock-shelters, some of respectable size. In *Early Man in Orange County, New York*, George R. Walters writes that "the Dutchess Quarry Cave is situated in the cliff at an elevation of 580 feet above sea level. The opening to the cave faces northwest. . . . The cave is roughly cylindrical in shape, 17 feet wide at the mouth and 60 feet long, narrowing down to a small fissure. . . . It evidently at one time carried a small subterranean stream." We presume this description is of either Cave #1 or Cave #8, which have proven to be the most significant archaeologically of the shelter caves explored.

History

Mount Lookout is a 580-foot-high hill that has been heavily quarried over the centuries.

The Dutchess Quarry Sites National Register District contains 13 acres of land on the western slope of Mount Lookout, including the summit. The Dutchess Quarry site contains four caves, 1, 2, 7, and 8; the Goshen Quarry Loci three caves, 3, 4, and 5. There are others still to be excavated in the Alpine Meadow section.

Orange County has owned the Mount Lookout property since the late 1830s, leasing out part of the mountain to the Goshen Quarry & Supply Company, which has been extracting dolomite for gravel.

In 1964, Native American artifacts were discovered at the cave site from hunter-gatherers dating back as far as 12,000 years ago. At the time of this discovery, it was the oldest site east of the Mississippi.

The cave site was added to the National Register of Historic Places in 1974.

Directions

From Finnegans Corners (junction of Routes 6/Pulaski Highway & 17A), drive southwest on Route 6/Pulaski Highway for 0.5 mile. Turn left onto Quarry Road and head south, which will put you in the general area where the shelter caves are located, to your left.

The site, being archaeologically sensitive, is not likely to be open to the public, except possibly for guided tours.

Resources

Patricia Edwards Clyne, *Hudson Valley Faces and Places* (Woodstock, NY: Overlook Press, 2005), 226–228.

"Dutchess Quarry Sites National Register District: Management and Conservation Report." https://www.orangecountygov.com/DocumentCenter/View/327/Dutchess-Quarry-Sites-National-Register-District---Management-and-Conservation-Report-2012-PDF.

"2011-06-11 Dutchess Quarry Cave Chapter Tour."https://ioccnysaa.blogspot.com/2011/06/2011-06-11-dutchess-quarry-cave-chapter.html.

R. E. Funk and D. W. Steadman, *Dutchess Quarry Caves, Orange County, New York* n.p.: Persimmon Press Monographs in Archaeology, 1994).

Persimmon Press Monographs in Archaeology, 1994).

Patricia Edwards Clyne, *Hudson Valley Trails and Tales* (Woodstock, NY: Overlook Press, 1990), 119.

George R. Walters, *Early Man in Orange County, New York* (Middletown, NY: Historical Society of Middletown and the Wallkill Precinct, 1973), 3.

Sullivan County

Sullivan County covers 997 square miles and is named after Major General John Sullivan, a Revolutionary War hero. Its highest point is 3,118-foot-high Beach Mountain near Hodge Pond.

Overhang Boulder 112

Long Path

Type of Formation: Large Boulder
WOW Factor: Not determined
Location: Summitville (Sullivan County)
Tenth Edition, NYS Atlas & Gazetteer: p. 102, D1
Earlier Edition, NYS Atlas & Gazetteer: p. 35, CD7
Parking GPS Coordinates: 41°36.665′N, 74°26.095′W
Overhang Boulder GPS Coordinates: Not determined
Accessibility: 0.7-mile hike
Degree of Difficulty: Moderate

Description

The only information we have about this boulder is taken from *The Long Path Guide*, which describes the rock as "a large overhanging boulder" with scenic views into the Roosa Gap and northwest to the Catskills.

Directions

From Ellenville (junction of Routes 209/Main Street & 52/Center Street), drive southwest on Route 209 for >8.0 miles. In Summitville, turn left onto Ferguson Road and head east for >0.1 mile. Then turn right, continuing on Ferguson Road/Rosa Gap-Summitville Road, and proceed southeast for 1.0 mile. Finally, bear left into a small parking area on your left.

Walk across the road and follow the Long Path south for 0.7 mile to reach the Overhang Boulder.

Resources

Herb Chong, ed., *The Long Path Guide*, 5th ed. (Mahwah, NJ: New York-New Jersey Trail Conference, 2002), 79.

Tri-States Monument and Witness Monument

113

Type of Formation: Historical Monument
WOW Factor: 3
Location: South of Port Jervis (Orange County)
Tenth Edition, NYS Atlas & Gazetteer: p. 106, B33
Earlier Edition, NYS Atlas & Gazetteer: p. 31, B5
Parking GPS Coordinates: 41°21.455'N, 74°41.657'W
Destinations GPS Coordinates: *Tri-States Monument*: 41°21.437'N, 74°41.684'W; *Witness Monument*: 41°21.445'N, 74°41.674'W
Accessibility: <0.1-mile walk
Degree of Difficulty: Easy

Description

The Tri-States Monument, aka Tri-State Rock, is a granite monument located at Carpenter's Point that approximately marks the junction of the state boundaries between New York, New Jersey, and Pennsylvania. We use the word "approximately" because the actual tri-state boundary is located 450 feet west of the Tri-States Monument, right in the middle of the Delaware River.

Even though the monument is land-based, it lies close enough to the river's high-water mark to be occasionally awash by water.

The Witness Monument, aka Reference Monument, bears witness to the boundaries of New York and Pennsylvania, but is physically on neither.

History

The Tri-States Monument is a fairly small, rectangular-shaped stone, measuring 2 feet high, and 1.4 feet wide, but it is very historical. Inscribed on

45. Tri-State Witness Monument. Antique postcard, public domain.

its upper surface are the initials of all three states, with groves to represent the state boundaries.

The Witness Monument is considerably taller than the Tri-States Monument and was erected in 1882. On one side of the monument is a reference to New York State; on the other side, to New Jersey.

Carpenter's Point is named for Benjamin Carpenter and his family who operated a ferry service across the Delaware River in this general area. The hamlet of Carpenter's Point was settled about 1690.

The Laurel Grove Cemetery was founded in 1856 by John Conkling. Today, nearly 14,000 people lay buried in what, to Victorians, was not only a final resting place for the dead, but a place where the living could come to enjoy solitude, picnic, or leisurely strolls.

Fittingly, John Conkling lies buried in the cemetery that he created.

Directions

From I-84 south of Port Jervis, take Exit 1 for Port Jervis/Sussex/Route 6/Route 23. If traveling west on I-84, turn left onto Route 6 and proceed northwest for <0.6 mile. If traveling east on I-84, turn right onto Route 15, go under I-84 and then up to Route 6 in 0.2 mile. Turn left onto Route 6 and proceed northeast for 0.4 mile.

Coming from either approach, turn left into the entrance for the Laurel Grove Cemetery [41°21.697′N, 74°41.148′W] and drive southwest for 0.5 mile until you are near the I-84 overpass.

Park to the side of the road. The Tri-State Monument and Witness Monument are both near the terminus of Carpenter's Point, which is defined by the confluence of the Delaware River and Neversink River. You will encounter the Witness Monument first. It is located under the east-bound lane of the I-84 overpass. The Tri-States Monument lies just a short distance beyond, near the water's edge.

Resources

Wikipedia, s.v. "Tri-States Monument," last modified February 11, 2022, https://en.wikipedia.org/wiki/Tri-States_Monument.

Minisink Battleground Park Rocks

114

Minisink Battleground Park

Type of Formation: Large Boulder; Rock-Shelter
WOW Factor: 4
Location: Barryville (Sullivan County)
Tenth Edition, NYS Atlas & Gazetteer: p. 105, A10
Earlier Edition, NYS Atlas & Gazetteer: p. 30, A3
Parking GPS Coordinates: 41°29.224'N, 74°58.193'W
Destination GPS Coordinates: *Minisink Spring Rock-Shelter*: Not taken; *Indian Rock*: 41°29.241'N, 74°58.141'W; *Hospital Rock*: 41°29.341'N, 74°58.211'W; *Sentinel Rock*: 41°29.334'N, 74°58.262'W; *Quarry*: 41°29.124'N, 74°58.181'W; *Minisink Battle Memorial*: 41°29.295'N, 74°58.230'W; *Minisink Battle Monument*: 41°29.291'N, 74°58.217'W
Accessibility: 0.9-mile-loop hike
Degree of Difficulty: Moderately easy
Additional Information: Minisink Battleground Park, Route 168, Barryville, NY 12719; (845) 807-0261

Description

The Minisink Spring Rock-Shelter consists of a large, overhanging ledge.

Hospital Rock is a ledge with a rock overhang.

Sentinel Rock is a 6-foot-high, 8-foot-long block of rock.

Indian Rock is a medium-sized boulder at the start of the Old Quarry Trail.

History

The Minisink Spring Rock-Shelter has been used by prehistoric people for as far back as 4,000 years.

It was here that the Battle of Minisink took place in 1779 when Joseph Brant, leading a band of Tories and Iroquois, vanquished nearly 50 New York and New Jersey militiamen.

Hospital Rock marks the spot where Lieutenant Colonel Benjamin Tusten and 17 wounded soldiers were trapped and killed by Joseph Brant's warriors.

Sentinel Rock is allegedly where one of the militiamen's guards was killed, but this account remains questionable.

A historic marker at Indian Rock states, "Legend has it that the Indians and Tories of Joseph Brant set this stone to honor their dead and wounded who fell before the field of fire from the nearby plateau."

The 57-acre battlefield was purchased in the early 1900s by the Minisink Valley Historical Society. Since 1955, the park has been maintained by the Sullivan County Parks and Recreation Commission.

Directions

From Barryville (junction of Routes 97 & 55), drive west on Route 97 for >4.0 miles. Turn right onto Route 168/Minisink Battleground Road and head northeast for 0.8 mile. At a fork, where York Lake Road goes right, bear left onto Mountain Road and proceed northwest for 0.1 mile to the entrance to the Minisink Battleground. The parking area is on your left.

From the parking area, follow the Battleground Trail northeast to reach Indian Rock. Then head northwest, turning right (east) at a junction onto the Woodland Trail. Very quickly, another trail leads off to your right and takes you up to the Minisink Spring Rock-Shelter.

Return to the Battleground Trail. Follow the trail north, then south, as it takes you on a large loop, first past Hospital Rock, and then over to Sentinel Rock. Eventually, you will return to the parking area.

If you wish to see the quarry and its rocks, follow the Old Quarry Trail, either from the Interpretive Center or from Indian Rock.

Resources

Lawrence C. Swayne, "Revolutionary War Heroism Acknowledged," *Kaatskill Life* 22, no. 1 (Spring 2007): 15–19. A photograph of Hospital Rock is shown on page 16.

"The Battle at Minisink Ford: July 22, 1779." http://minisink.org/minisinkbattle.html. This site includes a hand-drawn map of the battlefield.
"The Battle of Minisink." https://www.myrevolutionarywar.com/battles/790719-minisink.
"Minisink Battleground Park." https://scenesfromthetrail.com/2017/10/12/minisink-battleground-park.
Charles E. Stickney, *A History of the Minisink Region* (Middletown, NY: Coe Finch & I. F. Guiwits, 1867). The "Battle of Minisink" is recounted on pages 99–114.

Rockland County

Rockland County is the southernmost county in New York State, situated on the west side of the Hudson River. The name, meaning "rocky land," comes from the rock-strewn landscape, which includes the towering Palisades.

The county encompasses 199 square miles, of which only 26 square miles is water.

Nyack Balance Rock (Historic) 115

Type of Formation: Balanced Rock
Location: Blauvelt (Rockland County)
Tenth Edition, NYS Atlas & Gazetteer: p. 110, A5
Earlier Edition, NYS Atlas & Gazetteer: p. 32, D4
Parking GPS Coordinates: 41°03.739′N, 73°56.372′W
Nyack Balance Rock GPS Coordinates: 41°04.677′N, 73°55.820′W (estimated)
Accessibility: 2.0-mile hike to former site
Degree of Difficulty: Moderately difficult

46. Nyack Balanced Rock. Antique postcard, public domain.

Description

Postcard images of the Nyack Balance Rock show a boulder that is about 12 feet high and 20 feet long. In "Balance Rock and the Great South Mountain Raid," A. Touney writes, "The boulder itself was somewhere between 20 and 22 feet long, about 10 feet high at its thickest and 10 to 12 feet wide at its widest. Its northern end projected into the air five or six feet above the ground. And at its southern end, which was both narrower and lower than the rest of the rock, smaller rocks had been stacked to make it possible for the athletically inclined to climb onto its top surface and majestically observe the view."

Cornelia F. Bedell, in *Now and Then and Long Ago in Rockland County, New York*, writes that "probably the most perfect example of Pleistocene glacial boulders is Balance Rock on the crest of the hill southwest of the village of Nyack. It stands perfectly balanced and securely wedged on a high outcropping of rock well isolated from other rocks of similar form. . . . It is, however, a practically perfect example of a glacial rock ground to surprising symmetry."

Sadly, this grand boulder no longer exists.

History

According to Herb Chong, in *The Long Path Guide*, Balance Rock was removed by park officials in 1966 after vandals had destabilized it.

In *Now and Then and Long Ago in Rockland County, New York*, Cornelia F. Bedell writes that "Balance Rock is our greatest geological heritage"—words that come back to haunt us now that the rock is gone.

It's hard to believe that an historical rock of this importance, featured in numerous late nineteenth-century and early twentieth-century postcards, no longer exists because of human intervention. One can't help but wonder if more could have been done to stabilize and preserve the boulder.

WAS THERE A SECOND ROCK OF GEOLOGICAL NOTE?

Bedell mentions another rock close to Balance Rock:

> Not many years ago, Rockland County lost one of its great geological relics. Strangely enough, this second page of geological history rested only a few dozen yards from Balance Rock. The specimen in question was a large boulder bearing three or more glacial potholes—the span of

the hand in diameter—of varying depths. The inner surfaces of these depressions were highly polished by the action of swiftly flowing water and hard pebbles and grit. This was a rare specimen, indeed, but that did not save it, for unknown workmen on the old Tweed Road failed to recognize its value and so the relic was quickly blasted to bits.

From Bedell's account, it sounds like South Mountain not only lost one, but two unique, natural treasures.

Directions

From Central Nyack (junction of Routes 303 & 59), drive south on Route 303 for >1.8 miles. Turn left at a fork onto Greenbush Road and head southeast for 0.6 mile. Then turn left onto Clausland Mountain Road and drive northeast for <0.4 mile. Turn left into the parking area for Tackamack Town Park.

Follow the Long Path north for 2.0 miles to reach the site of the former Balance Rock. As far as we can determine, the boulder was not very far from Tweed Boulevard/Route 5. However, unless the site has been marked for its historical significance, you will probably hike right by it without notice.

Resources

Cornelia F. Bedell, *Now and Then and Long Ago in Rockland County, New York* (New City, NY: Historical Society of Rockland County, 1968), 10–11.

Herb Chong, ed., *The Long Path Guide*, 5th ed. (Mahwah, NJ: New York-New Jersey Trail Conference, 2002), 29.

A. Touney, "Local History Like You've Never Heard It: Balance Rock and the Great South Mountain Raid," *Patch* (Nyack-Piermont, NY), March 25, 2013, https://patch.com/new-york/nyack/bp--local-history-like-youve-never-heard-it. In the article, Touney gives a very detailed account of Balance Rock's history and ultimate demise. The author goes on to write, "Depending on its exact volume it would have weighed somewhere between 400,000 and 500,000 lbs. And to lift or roll it over the bedrock outcroppings would have required on the order of at least 2,000 strong teenagers rather than 20."

Monsey Glen Rockshelters 116

Monsey Glen County Park

Type of Formation: Rock-Shelter
WOW Factor: Not determined
Location: Monsey (Rockland County)
Tenth Edition, NYS Atlas & Gazetteer: p. 108, E1
Earlier Edition, NYS Atlas & Gazetteer: p. 32, CD3
Parking GPS Coordinates: 41°06.378′N, 74°04.223′W
Rock-shelter GPS Coordinates: Not determined
Hours: Daily, dawn to dusk
Accessibility: <0.3-mile hike
Degree of Difficulty: Moderately easy

Description

A number of rock-shelters that were used by Native Americans from about 1000 B.C. to 1600 A.D can be found in Monsey Glen. According to Patricia Edwards Clyne, in *Hudson Valley Trails and Tales*, "The largest of the glen's rockshelters was 49 feet long, 6 feet high, and 6 feet deep when archeologists first measured it in 1936." The rock-shelter was even larger at one time, as evidenced by a large sandstone slab lying on the ground that fell from the roof overhang many centuries ago.

Clyne writes that in the late 1950s, construction of the New York State Thruway destroyed much of the group's largest rock-shelter, but there are others nearby that remain preserved.

History

Monsey Glen is a 25-acre park containing sandstone overhangs that once served as shelters for early Native Americans.

The park land was acquired by the county in 1976. Artifacts from local Native Americans, including pottery, date as far back as 1,000 B.C. Some tools that were uncovered even deeper in the ground may be even more than 3,000 years old.

Directions

From Spring Valley (junction of Routes 59 & 45/South Main Street), drive west on Route 59 for 1.3 miles. Turn left onto Saddle River Road, head south for 0.1 mile, and then bear right into the parking area for Monsey Glen.

From the parking area, walk west to explore the park, which is no more than 0.3 mile in length.

The New York State Thruway is ever present and quite audible next to the south border of the park.

Resources

Patricia Edwards Clyne, *Hudson Valley Trails and Tales* (Woodstock, NY: Overlook Press, 1990), 68.

"Monsey Glen Park." https://nynjtc.org/park/monsey-glen-county-park.

"County Parks and Dog Runs." http://rocklandgov.com/departments/environmental-resources/county-parks-and-dog-runs/monsey-glen-park/.

Indian Rock and Spook Rock 117

Type of Formation: Large Rock; Historic Rock
WOW Factor: *Indian Rock*: 9; *Spook Rock*: 3
Location: Montebello (Rockland County)
Tenth Edition, NYS Atlas & Gazetteer: p. 107, E10
Earlier Edition, NYS Atlas & Gazetteer: p. 32, CD2–3
Destination GPS Coordinates: *Indian Rock*: 41°06.809′N, 74°07.748′W; *Spook Rock*: 41°07.178′N, 74°05.919′W
Accessibility: Roadside
Degree of Difficulty: Easy

Description

Indian Rock is a large, 18-foot x 9-foot x 15-foot granite boulder with an estimated weight of 17,300 tons. Geologists believe that the boulder, which formed 1.2 billion to 800 million years ago during the Precambrian era, was transported to its present location by glaciers from the nearby Ramapo Mountains-Hudson Highlands.

Although Indian Rock may look like a conglomeration of smaller rocks that have been somehow pushed together to form a pile, geologists believe that the rock was actually one single piece when dropped in place here by glaciers. Since then, stress fractures in the boulder, exploited by the repetitive effects of freezing and thawing, have caused it to break apart.

Spook Rock is a pile of rocks—a shattered boulder—much like Indian Rock.

History

Indian Rock was at one time part of the Native American route that led from upstate New York down to the rock, ending in Mahwah, New Jersey, where tribal meetings took place.

Indian Rock was nearly destroyed when Route 59 was being constructed. Engineers wanted to demolish the rock to make way for the highway. Fortunately, a "Save the Rock" movement pushed back against developers, the path of Route 59 was altered slightly, and Indian Rock was spared. Another source states that a developer wanted to create a shopping plaza by destroying the rock and was blocked from doing so by citizen action led by Craig Long. Either way, preservationists won out.

SPOOK ROCK

According to legend, human sacrifices were made on the rock altar. It is presumably the spirits of those tormented souls that remain in close proximity to the rock today. Another legend claims that a Dutch woman and her Native American lover were murdered near the rock by Dutch folks in what today would be called a hate crime.

Although Spook Rock is not as memorable when viewed as Indian Rock, its name still lives on at nearby businesses like the Spook Rock Senior Center and the Spook Rock Golf Course.

Directions

INDIAN ROCK

From Suffern (junction of Routes 59 & 89/South Airmont Road), drive west on Route 59 for <0.8 mile and turn into the parking area for the Indian Rock Shopping Center. The rock is located near the CVS Pharmacy.

From the New York State Thruway (I-87), take Exit 14B and drive south on North Airmont Road for 0.3 mile. Turn right onto Route 59, and head west for <0.8 mile to reach the Indian Rock Shopping Center.

SPOOK ROCK

From the New York State Thruway (I-87), take Exit 14 and head northeast on North Airmont Road for 0.8 mile to reach the junction of Spook Rock Road and North Airmont Road/Highview Road. Spook Rock is in view on top of a supporting stone wall at the southeast corner of the junction.

To get to Spook Rock from Indian Rock, head east on Route 59 for 0.8 mile. Turn left onto North Airmont Drive and proceed northeast for <1.2 miles to reach the junction with Spook Rock Road.

Resources

Linda Zimmerman, ed., *Rockland County Century of History* (New City, NY: Historical Society of Rockland County, 2002), 285. An article on the "Indian Rock Controversy" includes a picture of Indian Rock when it was photographed in 1985 with trees around it, as well as a photograph of the rock in 2001 in the shopping plaza.

"Paranormal Investigations of Rockland County Local Lore and Legends." https:/ pirc-ny.com/local-lores and legends.

Cornelia F. Bedell, *Now and Then and Long Ago in Rockland County* (Rockland County, NY: Privately printed, 1941), 33.

Dater Mountain Erratic 118

Dater Mountain Nature Park

Type of Formation: Large Boulder
WOW Factor: 6
Location: Sloatsburg (Rockland County)
Tenth Edition, NYS Atlas & Gazetteer: p. 107, DE9
Earlier Edition, NYS Atlas & Gazetteer: p. 32, C2
Parking GPS Coordinates: 41°10.345′N, 74°10.533′W
Dater Mountain Erratic GPS Coordinates: 41°10.048′N, 74°11.072′W
Accessibility: 1.0-mile hike
Degree of Difficulty: Moderate
Dater Mountain Nature Park Trail Map: rocklandgov.com/files/4913/4555/8821/Parks_Dater_Mountain_Map.pdf

47. Dater Mountain Erratic. Photograph by Dan Balogh.

Description

The Dater Mountain glacial erratic is 10 feet high and lumpy looking, lying in close proximity to a line of telephone poles overlooking the New York State Thruway.

History

The Dater Mountain Nature Park encompasses 350 acres of land that were acquired in two stages between 1981 and 2004.

During the late eighteenth century into the nineteenth century, the trees on this hilly tract of land were extensively harvested to produce charcoal for the smelting of iron ore. Many of these charcoal pits still dot the landscape.

Dater Mountain, Dater Mine, and by logical extension the Dater Mountain Erratic, are possibly named for Abram Dater or one of his ancestors.

Directions

From Sloatsburg (junction of Seven Lakes Drive and Route 17), drive northeast on Seven Lakes Drive for 0.5 mile. As soon as you pass under I-87/NYS Thruway, turn left onto Johnsontown Road and head northeast for 0.8 mile. Park to your right in a small pull-off.

Walk across the road and follow the orange-blazed trail uphill. When you come to a junction after 0.1 mile, turn left onto the blue-blazed trail and proceed southwest. After you have climbed up Sleater Hill and partially descended, look for a short path/road to your left that leads to a large glacial erratic by a set of telephone poles, not that far above the New York State Thruway.

Resources

"Dater Mountain Nature Park." https://www.nynjtc.org/hike/dater-mountain-nature-park.

"Dater Mountain Park, Johnsontown Road, Sloatsburg." http://rocklandgov.com/departments/environmental-resources/county-parks-and-dog-runs/dater-mountain-park.

High Tor and Little Tor Erratics 119

High Tor State Park

Type of Formation: Large Boulder
WOW Factor: 5
Location: West of Haverstraw (Rockland County)
Tenth Edition, NYS Atlas & Gazetteer: p. 108, D1–2
Earlier Edition, NYS Atlas & Gazetteer: p. 32, C3–4
Parking GPS Coordinates: *High Tor State Park*: 41°11.478'N, 73°59.371'W; *Long Path*: 41°11.528'N, 74°00.271'W
Destination GPS Coordinates: *High Tor Boulders*: Not determined; *Little Tor Erratic*: 41°11.780'N, 73°59.275'W (Google Earth)
Fee: *High Tor State Park*: Admission charged; *Long Path*: no charge
Hours: Check High Tor State Park website for specific details
Accessibility: 2.0-mile hike (estimated)
Degree of Difficulty: Moderately difficult
Additional Information: High Tor State Park, 415 South Mountain Road, New City, NY 10956
Trail Map: parks.ny.gov/documents/parks/HighTorTrailMap.pdf

Description

According to John Serrao in *The Wild Palisades of the Hudson*, "several large granite and gneiss erratics were . . . dropped along ridge by glaciers, and can be seen off the Long Path."

One large erratic rests on top of Little Tor, and the GPS coordinates for it were taken off a listing on Google Earth. Raymond H. Torrey, Frank Place Jr., and Robert L. Dickinson, in *New York Walk Book*, write, "Beyond this, glimpses can be had of Little Tor, identified by the gleaming white boulder perched on its summit."

In *Now and Then and Long Ago in Rockland County New York*, Cornelia F. Bedell describes the view from the mountain summit: "we look up along the mountain ridge which extends northward along the river. There, not many yards away, is the first of several huge boulders resting on the very brink of the cliff. These huge boulders were carried along by the glacier."

History

High Tor State Park contains two significant peaks on South Mountain—797-foot-high High Tor, and 620-foot-high Little Tor. Although these peaks may not reach the elevation of the high peaks in the Catskills and the High Peaks in the Adirondacks, they are the two highest points in the Hudson Palisades.

Little Tor has the distinction of having been used by colonists as a signal point during the American Revolutionary War. The property was acquired by the Palisades Interstate Park Commission in 1943.

The Long Path crosses through High Tor State Park, following along the ridge line for 3.5 miles.

There seems to be two possible definitions of the word *Tor*. One source states that the word is English for "rocky peak," another that it is a Celtic word for "gateway," a place to commune with the gods.

Directions

HIGH TOR STATE PARK ENTRANCE

From north of West Nyack (junction of Routes 9W & 303/Country Ridge Road), drive northwest on Route 9 for 0.6 mile. At a traffic intersection, turn left onto Route 304 and proceed southwest for 1.1 miles. When you come to Ridge Road, turn right and head north for 0.8 mile. Then turn left onto South Mountain Road/Haverstraw Road and head northwest for 1.6 miles. Finally, bear right into the High Tor State Park entrance and drive north for >0.2 mile to arrive at the parking area. From the parking area, follow the park road northeast to Tam's Point and, from there, the <0.2-mile long Tam's Path north up to the Long Trail. From this point, you can continue north on the Small Tor Spur Path up to the summit of Little Tor.

High Tor is roughly 1.3 miles southeast of Little Tor.

LONG PATH ENTRANCE

Driving past the entrance to High Tor State Park, continue west on South Mountain Road for another 1.0 mile. Turn right onto Route 33/South Central Highway and proceed north for 0.8 mile. Park in a small pull-off to your right. From here, pick up the Long Path, which takes you east, then south, as it follows along the High Tor ridgeline.

Resources

John Serrao, *The Wild Palisades of the Hudson* (Westwood, NJ: Lind Publications, 1949), 150.

Barbara H. Gottlock and Wesley Gottlock, *New York's Palisades Interstate Park*, Images of America Series (Charleston, SC: Arcadia Press, 2007), 107. A photograph shows the summit of High Tor with a tower that was once used by pilots for navigational purposes. The tower was removed in 1963.

Raymond H. Torrey, Frank Place Jr., and Robert L. Dickinson, *New York Walk Book*, 3rd ed. (New York: American Geographical Society, 1951), 146.

Cornelia F. Bedell, *Now and Then and Long Ago in Rockland County* (New City, NY: Historical Society of Rockland County, 1968), 10.

André the Spy Rock 120

Haverstraw Beach State Park

Type of Formation: Historic Rock
WOW Factor: 3
Location: Haverstraw (Rockland County)
Tenth Edition, NYS Atlas & Gazetteer: p. 108, D2
Earlier Edition, NYS Atlas & Gazetteer: p. 32, C4
Parking GPS Coordinates: 41°10.753′N, 73°56.739′W
Rocky Point GPS Coordinates: 41°10.583′N, 73°56.344′W
André the Spy Rock GPS Coordinates: 41°10.539′N, 73°56.339′W (estimated); 41°10.667′ N, 73°56.582′ W (Historical Marker Database website)
Accessibility: 0.3–0.4-mile hike
Degree of Difficulty: Moderately easy
Additional Information: 200 Riverside Avenue, Haverstraw, NY 10927

Description

André the Spy Rock is a medium-sized, granite boulder that rests along the Hudson River shoreline. The rock supposedly marks the spot where General Benedict Arnold, turning traitor in 1780, exchanged the plans to West Point's defenses with John André, a British spy. The words "André the spy landed here Sept. 1780" have been etched into the rock.

History

Haverstraw Beach State Park is a component of the Palisades Interstate Park system. The 73-acre, minimally developed Haverstraw Beach State Park was established in 1911, occupying land that was previously known as Snedeker's Landing and Waldbery Landing.

For those who are curious, there is also a Major John André Monument on André Hill Road in Tappan, NY (41°01.281′N, 73°57.289′W), which commemorates the site of Major André's hanging.

Directions

From north of West Nyack (junction of Routes 9W & 303/County Ridge Road), drive northwest on Route 9W for 2.0 miles. Turn right onto Short Clove Road and, after 0.3 mile, bear left onto Riverside Avenue and head southeast for 0.9 mile, parking in an area at the end of the road.

Walk past the yellow barrier and proceed south along the continuation of Riverside Avenue for 0.4 mile. Then follow a path left for 0.1 mile to the arrowhead-shaped point of land on Haverstraw Beach called Rocky Point.

According to directions provided by Raymond H. Torrey, Frank Place Jr. and Robert L. Dickinson, in *New York Walk Book*, "walk downstream a few hundred feet" from the rocky point to reach the boulder—a walk of roughly 0.1 mile southeast from the point. You will come to several mid-sized boulders, one of which is André the Spy Rock.

Resources

Raymond H. Torrey, Frank Place Jr., and Robert L. Dickinson, *New York Walk Book*, 3rd ed. (New York: American Geographical Society, 1951), 141.

Cornelia F. Bedell, *Now and Then and Long Ago in Rockland County* (New City, NY: Historical Society of Rockland County, 1968), 373. Photograph of the André the Spy monument by Richard J. Kale.

"André the Spy Historical Marker." https://www.hmdb.org/m.asp?m=214383.

Harriman State Park and the Palisades

We are indebted to Dan Balogh and his wife, Laura Petersen Balogh, who have contributed the majority of the photographs that accompany the Harriman and Palisades section of this book. The Baloghs are lifetime residents of New Jersey who have been hiking trails in New York and New Jersey for the past 25 years. Over time, they have created a photo-journal of all of their outings that can be viewed on Dan's website, danbalogh.com.

Harriman State Park: Northern Section

History

Harriman State Park, encompassing 47,527 acres (75 square miles) of land, contains over 200 miles of hiking trails and 31 lakes and ponds, and spills over into both Orange and Rockland Counties. It is the second largest park in New York State's park system.

Harriman State Park is named for Edward and Mary Averell Harriman, who donated the initial 10,000 acres of land out of which the park grew. The state took possession of the land in 1910, and the Palisades Interstate Park Commission, beginning in 1913, began constructing dozens of group campsites throughout the park.

There are seemingly an infinite number of glacial erratics in Harriman State Park. The ones described here are those that have been given, in some instances, colorful names, and have become notable destinations or waypoints along hikes. The majority of the rocks tend to be trailside.

If you are interested in knowing about all the features on the Harriman Trail maps, consult the Map Index: Harriman-Bear Mountain Trails Map Set at nynjtc.org/content/index-harriman-bear-mountain-trails-map.

Resources

"Harriman State Park." https://parks.ny.gov/parks/145/details.aspx.

Stockbridge Cave Shelter and Hippo Rock

121

Harriman State Park

Type of Formation: Large Boulder; Shelter Cave
WOW Factor: 5
Location: Harriman (Rockland County)
Tenth Edition, NYS Atlas & Gazetteer: p. 108, C1
Earlier Edition, NYS Atlas & Gazetteer: p. 32, B3
Parking GPS Coordinates: 41°18.934′N, 74°02.993′W
Destination GPS Coordinates: *Stockbridge Cave Shelter*: 41°18.092′N, 74°04.620′W (estimated); *Stockbridge Shelter*: 41°18.101′N, 74°04.662′W (Google Earth); *Hippo Rock*: 41°17.870′N, 74°04.793′W (estimated)
Accessibility: *Stockbridge Cave Shelter*: 1.7-mile hike; *Hippo Rock*: 2.2-mile hike
Degree of Difficulty: Moderate
Additional Information: Northern Harriman Bear Mountain Trails, Trail Map 4
Trail Map: nynjtc.org/map/bear-mountain-day-hikes-map

Description

Stockbridge Cave Shelter is an intriguing rock-shelter that hikers have augmented, using stones to create an artificial chimney. Hershel Friedman, in his blog, *The Harriman Hiker*, writes, "The Cave Shelter is a rock formation of massive boulders that are stacked and forming caves and passageways that you can actually squeeze through quite deeply."

Hippo Rock is an enormous, cantilevered rock, some 8 feet high and decidedly longer, which overhangs a ledge. It seemingly defies gravity because of its greater mass resting on solid ground.

48. Stockbridge Cave Shelter. Photograph by Dan Balogh.

History

The Stockbridge Cave Shelter was "discovered" by J. Ashton Allis in 1922. It became known as the Stockbridge Cave Shelter in 1928 as park authorities tried to make its interior more hospitable. Today, the shelter is reputedly used by hikers as a backup for refuge when Stockbridge Shelter—a rock-and-mortar structure with a plank floor—is full.

According to *The Long Path Guide*, "The Cave Shelter is set into an overhang near the base of its rock face. It is damp and hardly an inviting place to spend the night." Still, something is better than nothing if it is shelter that you are seeking.

The name Stockbridge, which also applies to a nearby 1,320-foot-high mountain, comes from Elisha Stockbridge, who owned a hotel near Summit Lake.

The Stockbridge Shelter, passed along the way between the Stockbridge Cave Shelter and Hippo Rock on an area of exposed bedrock, was built in 1928 and is a fairly large structure.

Hippo Rock received its name from an imaginative hiker who looked at the rock and saw a hippopotamus.

49. Hippo Rock. Photograph by Dan Balogh.

Directions

From northwest of Harriman (junction of Routes 6/Long Mountain Parkway & 293), drive west on Route 6/Long Mountain Parkway for 2.0 miles, passing by a road exit at 1.8 miles, and turn left where a sign reads "Long Path. Raymond H. Torrey Memorial."

From the Seven Lakes Drive & Route 6 roundabout, i.e., Long Mountain Circle, drive northwest on Route 6 for 1.2 miles and turn right where a sign reads "Long Path. Raymond H. Torrey Memorial."

From either approach, park along the side of the road near the north end of the semicircular park road. Walk across Route 6 and follow the aqua-blazed Long Path southwest for 1.7 miles. Look for the Stockbridge Cave Shelter to your left, distinguished by a stone chimney. At 0.3 mile from the Stockbridge Cave Shelter, you will pass by the more hospitable Stockbridge Shelter.

At 2.2 miles, you will come to Hippo Rock on your right, <0.2 mile southwest of where the yellow-blazed Menomine Trail crosses the Long Path. If you continue southwest on the Long Path for another 2.0 miles, you will cross Arden Valley Road (41°16.663′N, 74°05.771′W), where there is no parking except at Lake Torati Beach (41°16.515'N, 74°05.335'W), 0.5 mile east.

Resources

Jerome Wyckoff, *Rock Scenery of the Hudson Highlands and Palisades* (Glens Falls, NY: Adirondack Mountain Club, 1971), 85.

New York-New Jersey Trail Conference, *New York Walk Book*, 6th ed. (New York: New York-New Jersey Trail Conference, 1998). Mention is made of Hippo Rock on page 39.

William J. Myles and Daniel Chazin, *Harriman Trails: A Guide and History*, 4th ed. (New York: New York-New Jersey Trail Conference, 2018), 63. Photograph of Hippo Rock.

Herb Chong, ed., *The Long Path Guide*, 5th ed. (Mahwah, NJ: New York-New Jersey Trail Conference, 2002), 52.

Leonard M. Adkins and the Appalachian Trail Conservancy, *Along the Appalachian Trail, New Jersey, New York, and Connecticut*, Images of America Series (Charleston, SC: Arcadia Publishing, 2014). A photograph of the Stockbridge Cave Shelter is shown on page 52.

"The Lonesome Hiker: Stockbridge Shelter Cave, Cave Shelter, Hippo Rock – Harriman State Park, NY." https://lonehiker.blogspot.com/2009/04/stockbridge-shelter-cave-shelter-hippo.html.

The Timp and Collar Button

122

Type of Formation: Boulder: Rock Profile
WOW Factor: 6
Location: Jones Point (Rockland County)
Tenth Edition, NYS Atlas & Gazetteer: p. 108, C2
Earlier Edition, NYS Atlas & Gazetteer: p. 32, BC4
1777 Trailhead Parking GPS Coordinates: 41°16.438′N, 73°58.387′W
Destination GPS Coordinates: *The Timp*: Not determined; *Collar Button*: Not determined
Accessibility: 3.0–4.0-mile hike depending upon route taken
Degree of Difficulty: Difficult
Additional Information: Northern Harriman Bear Mountain Trails, Trail Map 4
Trail Map: parks.ny.gov/documents/parks/BearMountainTrailMap.pdf

---

Description

In the *New York Walk Book*, Raymond H. Torrey, Frank Place Jr., and Robert L. Dickinson write, "Above frowns the Timp, a striking cliff with a pronounced overhang, at the west end of the Dunderberg massif."

William J. Myles, in *Harriman Trails: A Guide and History*, writes about a "large boulder sitting in the [Timp] pass" called Collar Button. One would imagine the rock assuming the shape of a button.

A number of interesting boulders are also encountered along the hike, which to the best of our knowledge remain nameless. Here's your chance to have a rock named for you.

History

The 1777 Trail follows the route taken by British soldiers under the leadership of Sir Henry Clinton as they marched from Stony Point in 1777 to attack Fort Clinton and Fort Montgomery.

50. Hiking along the Timp Trail. Photograph by Dan Balogh.

The Timp is a large, 1,080-foot-high mountain with a distinctive face. Timp Pass is a deep notch between the Timp and 1,250-foot-high West Mountain.

Timp is an obsolete Dutch word for a loaf of bread or cake. Perhaps some early Dutch explorers likened the shape of the mountain to that of a loaf of bread.

Directions

From Tompkins Cove (junction of Routes 9W/202/North Liberty Drive & 118/Mott Farm Road), drive northeast on Route 9W/202/North Liberty Drive. After 1.2 miles, turn right into a parking area.

From the parking area, walk north along Route 9W for around 100 feet. Cross over to the west side of the road and pick up the red-blazed 1777 Trail. Proceed west for >1.3 miles. At a junction, turn left onto the blue-blazed Timp-Torne Trail and continue west for 0.5 mile. When you come to the next junction, turn left onto the red-blazed Ramapo-Dunderberg Trail, which takes you south, then northwest as you go through the Timp Pass.

The Collar Button Boulder is close to the junction of the red-blazed Red Cross Trail (coming in from the south), the unmarked Timp Pass

Road (coming in from the north), and the Ramapo-Dunderberg Trail (which you are on).

To turn the hike into a loop, follow the Timp Pass Road north for a short distance, and then right onto the blue-blazed Timp-Torne Trail as it takes you east up and over the Timp, and eventually down to the red-blazed 1777 Trail, and back east to the parking area.

Resources

Raymond H. Torrey, Frank Place Jr., and Robert L. Dickinson, *New York Walk Book*, 3rd ed. (New York: American Geographical Society, 1951), 176. On page 156, the meaning of the word *timp* is discussed.

"Best Hikes in Harriman State Park #1: The Timp and West Mountain." https://www.myharriman.com/best-hikes-in-harriman-1-the-timp-and-west-mountain.

William J. Myles, *Harriman Trails: A Guide and History* (New York: The New York-New Jersey Trail Conference, 1994), 156.

Cape Horn Rock and Cat's Den 123

Type of Formation: Large Rock
WOW Factor: 4–5
Location: Lake Skannatati (Orange County)
Tenth Edition, NYS Atlas & Gazetteer: p. 107, D10
Earlier Edition, NYS Atlas & Gazetteer: p. 32, BC2–3
Parking GPS Coordinates: 41°14.517′N, 74°06.138′W
Cape Horn Rock & Cat's Den GPS Coordinates: Not determined
Accessibility: 1.4-mile hike
Degree of Difficulty: Moderate
Additional Information: Northern Harriman Bear Mountain Trails, Trail Map 4.
Trail Map: parks.ny.gov/documents/parks/HarrimanTrailMap.pdf

Description

According to *The Long Path Guide*, Cape Horn Rock is an "overhanging rock" near where the trail turns right. William J. Myles, in *Harriman Trails: A Guide and History*, describes it as "a great boulder" and also mentions that there is a cave in the rocks behind the boulder, presumably a talus cave, that is called "Cat's Den."

History

Lake Skannatati, from where the hike begins, is a 36-acre body of water.

Seven Lakes Drive, a main road through the park, was originally called Southfields Road.

Directions

From Sloatsburg (junction of Seven Lakes Drive & Route 17), drive north-east on Seven Lakes Drive for nearly 8.0 miles (or 0.8 mile north past the Kanawauke Circle). Turn sharply left at the sign that reads "Welcome to

51. Cape Horn. Photograph by Dan Balogh.

Lakes Askoti & Skannatati. Parking" and park in the medium-sized lot in front of Lake Skannatati.

From the north end of the parking area, follow the aqua-blazed Long Path west, initially along the north shore of Lake Skannatati, for 1.4 miles to reach Cape Horn Rock. Along the way, the Dunning Trail will enter on your right at <1.3 miles and then exit on your left at >1.3 miles.

Resources

Herb Chong, ed., *The Long Path Guide*, 5th ed. (Mahwah, NJ: New York-New Jersey Trail Conference, 2002), 50.

"Lake Skannatati to US Route 6." https://www.nynjtc.org/book/6-lake-skannatati-us-route-6.

"Hogencamp Mountain." https://www.hikingproject.com/trail/7047052/hogencamp-mountain. Cape Horn is described as "a giant rock formation."

William J. Myles, *Harriman Trails: A Guide and History* (New York: New York-New Jersey Trail Conference, 1994), 69.

Rock House 124

Type of Formation: Rock-Shelter
WOW Factor: Unknown
Location: Kanawauke Circle (Orange County)
Tenth Edition, NYS Atlas & Gazetteer: p. 107, D10
Earlier Edition, NYS Atlas & Gazetteer: p. 32, BC2–3
Parking GPS Coordinates: 41°14.150′N, 74°05.956′W
Rock House GPS Coordinates: 41°14.158′N, 74°05.990′W (Estimated)
Accessibility: 0.05-mile-hike
Degree of Difficulty: Moderate
Trail Maps: Northern Harriman Bear Mountain Trails, Trail Map 4.
Southern Harriman Bear Mountain Trails, Trail Map 3
parks.ny.gov/documents/parks/HarrimanTrailMap.pdf

Description

The Rock House is either a rock-shelter formed by an overhanging ledge or a grouping of rocks that have come together to form a rock-enclosed shelter. William J. Myles in *Harriman Trails: A Guide and History* simply calls it a "cave."

History

Rockhouse Mountain (1,283 feet), the highest point in Rockland County, was named for the Rock House located on its shoulder.

"Kanawauke" (Kanawauke Circle) comes from the Onondaga word *kanawahkee,* which means "place of much water." This definition aptly applies to Lake Kanawauke, formerly known as Little Long Lake, which is a sizeable body of water. Unlike some of the other lakes in Harriman State Park, it is naturally formed.

Directions

From Kanawauke Circle (junction of Route 106/Kanawauke Road & Sevens Lakes Drive), drive east on Route 106/Kanawauke Road for 0.7 mile and turn right into a tiny pull-off, approximately 0.2 mile east from where the Long Path crosses Route 106/Kanawauke.

Walk into the woods, heading south downhill from the road. You may have to do a little scouting, but the Rock House should be encountered within a couple of hundred feet according to the Southern Harriman Bear Mountain Trails (Trail Map 3), which clearly shows the rock's position. The location is also noted in *Harriman Trails: A Guide and History*, where the authors write, "The 'rock house,' a cave which gives its name to the mountain, is below (south of) the highway, about 0.2 mile east of the Long Path crossing."

Resources

William J. Myles and Daniel Chazin, *Harriman Trails: A Guide and History*, 4th ed. (New York: New York-New Jersey Trail Conference, 2018), 241.

Bowling Rocks, Ship Rock, Times Square Boulder, Plateau Rocks, and Unnamed Pothole

125

Type of Formation: Large Boulder; Group of small rocks; pothole
WOW Factor: 3–8
Location: Kanawauke Circle (Orange County)
Tenth Edition, NYS Atlas & Gazetteer: p. 107, D10
Earlier Edition, NYS Atlas & Gazetteer: p. 32, BC2–3
Parking GPS Coordinates: 41°13.816′N, 74°08.413′W
Destination GPS Coordinates: *Bowling Rocks*: 41°14.647′N, 74°07.572′W (Google Earth); *Ship Rock:* 41°14.954′N, 74°07.646′W (Google Earth); *Times Square Boulder*: 41°15.113′N, 74°07.204′W (Google Earth); *Unnamed pothole*: Not taken; *Plateau Rocks*: 41°14.949′N, 74°07.265′W (estimated)
Accessibility: *Bowling Rocks*: 1.9-mile hike; *Ship Rock*: 5.0-mile hike; *Times Square Boulder*: 5.7-mile hike; *pothole*: 5.8-mile hike
Degree of Difficulty: Moderately difficult
Trail Maps: Northern Harriman Bear Mountain Trails, Map 4
Southern Harriman Bear Mountain Trails, Map 3
parks.ny.gov/documents/parks/HarrimanTrailMap.pdf

Description

Bowling Rocks consist of a smattering of stones scattered every which way on top of exposed bedrock, or, as William J. Myles in *Harriman Trails: A Guide and History* writes, "a level area of bare rock dotted by boulders." The roundness of the rocks is what prompted some creative hiker to come up with the bowling balls analogy.

52. Times Square Boulder. Photograph by Dan Balogh.

Ship Rock is a large, 25-foot-high, fractured boulder that, to some, with its raised bottom, resembles a ship's prow. The trail passes directly in front of the rock.

New York State also has a water-based Ship Rock in Rye that Richard Lederer writes about in *The Place-Names of Westchester County, New York*: "The rock in Mamaroneck Harbor, south of Hen Island and east of Crab Island, is submerged and marked by a Coast Guard buoy." A number of ships have been damaged or sunk by this rock, which is how the rock acquired its name through association. Google Earth lists Ship Rock at a GPS coordinate of 40°55.905′N, 73°42.558′W.

Times Square Boulder is justifiably named: first, the boulder sits at the busy intersection of five trails—the Ramapo-Dunderberg Trail, the Arden-Surebridge Trail (A-SB), and, 75 feet away, the Long Path; and secondly, somebody painted the words "Times Square" in big, white letters on the boulder, including an arrow that points the way toward New York City.

Times Square Boulder stands 6–8 feet high and 10–12-feet long, with a tiny shelter under one end. Near the rock is a stone fireplace. It is a favorite meeting place for hikers approaching from various directions.

The unnamed pothole, located on a ledge to the left of the trail, was "discovered" in 1924 by Frank Place. Peggy Turco, in *Walks and Rambles in the Western Hudson Valley*, describes it as "a chimney-like half cylinder 4 feet wide and 8 feet deep." Turco goes on to comment that "This is a glacial pothole. Usually, potholes are formed by streams where gravel whirls for millennia scouring a round pot-like depression in the stream's bedrock. The lack of any streambed at this site precludes that typical formation. It has been postulated that a stream once ran on the Wisconsin glacier. As the meltwater coursed into a crevasse at this spot, it scoured out the pothole." This is a very interesting hypothesis.

The potholes are also described by Raymond H. Torrey, Frank Place Jr. and Robert L. Dickinson, in *New York Walk Book*, who note that the path goes "around a low cliff and beneath a glacial pothole, 6 or 7 feet high on its face."

Plateau Rocks consist of a number of small-to-medium-sized boulders scattered across an area of exposed bedrock at the summit of Hogencamp Mountain (41°14.955'N, 74°07.262'W). Also visible is an oblong-shaped boulder propped up on one end by a tiny rock.

53. Ship Rock. Photograph by Dan Balogh.

Directions

From the Kanawauke Circle (junction of Route 106/Kanawauke Road & Seven Lakes Drive), drive west on Route 106/Kanawauke Road for 1.6 miles (or <0.3 mile past Masonic Camp Road). Park to your right.

Hike northeast on the red-blazed Ramapo-Dunderberg Trail for 1.5 miles. Along the way, you will pass the Nurian Trail on your left at 0.5 mile. At the junction with the yellow-blazed Dunning Trail, turn right and go 0.4 mile to reach Bowling Rocks, on your left.

Return to the Ramapo-Dunderberg Trail and continue north. After passing by the blue-blazed Lichen Trail at <2.0 miles, you will see Ship Rock within <0.05 mile from the junction.

Continue east, then north, on the Ramapo-Dunderberg Trail for another 0.7 mile. Just before the junction with the Long Path/Arden-Surebridge Trail at >2.6 miles, Times Square rock will appear on your right.

A notable pothole contained on a ledge is visible to your left along the Ramapo-Dunderberg Trail, 0.1 mile past the Long Path/Arden-Surebridge Trail intersection.

OTHER APPROACHES TO TIMES SQUARE

From the Lemon Squeezer (see next chapter), proceed southeast on the red-blazed Arden-Surebridge Trail/Long Path for 1.1 miles to reach the Times Square junction (or a total of 4.6 miles on the Arden-Sturbridge Trail from Elk Pen).

Alternatively, from the north end of the parking area for Lake Skannatati, follow the Long Path west, initially along the north shore of Lake Skannatati, then north, for 2.0 miles to reach the Times Square junction.

Resources

Jerome Wyckoff, *Rock Scenery of the Hudson Highlands and Palisades* (Glens Falls, NY: Adirondack Mountain Club, 1971). Photos of some medium-sized boulders on a scoured plateau of bedrock are shown on pages 13 and 83.

"Bald Rocks Hogencamp Mountain Hike" https://www.myharriman.com/bald-rocks-hogencamp-mountain-hike-harriman-state-park. A photograph of the Times Square Boulder is shown on the website.

Peggy Turco, *Walks and Rambles in the Western Hudson Valley* (Woodstock, VT: Backcountry Publications, 1996), 39.

Raymond H. Torrey, Frank Place Jr., and Robert L. Dickinson, *New York Walk Book*, 3rd ed. (New York: American Geographical Society, 1951), 175.

William J. Myles and Daniel Chazin, *Harriman Trails: A Guide and History*, 4th ed. (New York: The New York-New Jersey Trail Conference, 2018), 19–20. (Times Square Boulder).

Richard M. Lederer Jr., *The Place-Names of Westchester County, New York* (Harrison, NY: Harbor Hill Books, 1978), 131.

"Fingerboard Mountain Loop from Lake Skannatati – Harriman State Park." https://scenesfromthetrail.com/2017/10/28/fingerboard-mountain-loop-from-lake-skannatati/.

Lemon Squeezer 126

Type of Formation: Narrow passageway
WOW Factor: 8
Location: Southfields (Orange County)
Tenth Edition, NYS Atlas & Gazetteer: p. 107, CD10
Earlier Edition, NYS Atlas & Gazetteer: p. 32, BC2
Parking GPS Coordinates: 41°15.887′N, 74°09.260′W
Lemon Squeezer GPS Coordinates: 41°15.487′N, 74°07.989′W (Google Earth)
Accessibility: 2.0-mile hike
Degree of Difficulty: Difficult
Trail Maps: Northern Harriman Bear Mountain Trails, Trail Map 4
parks.ny.gov/documents/parks/HarrimanTrailMap.pdf

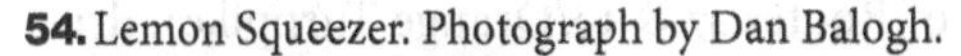

54. Lemon Squeezer. Photograph by Dan Balogh.

Description

The Lemon Squeezer, as its name suggests, is a very tight passageway between two large sections of rock. In *Harriman Trails: A Guide and History*, William J. Myles states that the Lemon Squeezer is "a narrow passage formed when a great piece of rock broke off the side of the cliff." Pictures of the Lemon Squeezer show a passageway 3 feet high, 1 foot wide, and angled at 15°. It is just one part of a massive area of rock lined with pine trees that you make your way through.

History

The Lemon Squeezer was named by J. Ashton Allis, one of the founders of the Palisades Interstate Conference.

The peculiar name of the parking area, Elk Pen, came about from an experiment that was tried in 1919. Sixty elk were brought over from Yellowstone National Park and kept in a wired enclosure between Arden and Southfields. The elk failed to thrive, however, and the attempt to introduce this species into the region ended in the 1940s. The name, Elk Pen, however, has continued on.

Directions

From Southfields (junction of Routes 17 & 19/Orange Turnpike), drive northeast on Route 17 for 1.9 miles. Turn right onto Arden Valley Road and proceed east for 0.3 mile, crossing over the New York State Thruway in the process. Then bear right onto a short road that takes you to the Elk Pen parking area in <0.1 mile.

Follow the white-blazed Appalachian Trail east for 3.4 miles. Along the way, you will cross over the outlet stream from Island Pond at 1.7 miles.

The Lemon Squeezer is reached at 2.0 miles near the junction with the red-blazed Arden Surebridge Trail. If you don't wish to go through the Lemon Squeezer, follow a blue-blazed side path that stouter or less adventurous hikers use to bypass the narrow squeeze.

If you continue east on the Arden Surebridge Trail as the Appalachian Trail goes north, you will come to the Long Path at 2.7 miles.

Resources

Raymond H. Torrey, Frank Place Jr., and Robert L. Dickinson, *New York Walk Book*, 3rd ed. (New York: American Geographical Society, 1951), 168. The Lemon Squeezer is portrayed as "a cliff and huge tumbled rock fragments."

"Harriman State Park, NY: Island Pond and Lemon Squeezer" https://gonehikin.blogspot.com/2012/02/harriman-state-park-island-pond-and.html. The blog contains a photograph of the lemon squeezer.

William J. Myles, *Harriman Trails: A Guide and History* (New York: The New York-New Jersey Trail Conference, 1994), 14. A photograph of the Lemon Squeezer is shown on page 13.

"Harriman State Park: Lemon Squeezer to Lichen Trail." https://hikethehudson-valley.com/hikes/harriman-state-park-lemon-squeezer-to-lichen-trail.

Peggy Turco, *Walks and Rambles in the Western Hudson Valley* (Woodstock, VT: Backcountry Publications, 1996), 41. The Lemon Squeezer is described as a "narrow cleft in the granite cliffs."

New York-New Jersey Trail Conference. *Guide to the Appalachian Trail in New York and New Jersey*. 9th ed. (Harpers Ferry, WV: Appalachian Trail Conference, 1983), 84. The Lemon Squeezer is depicted as "a narrow, steep passage between boulders."

Leonard M. Adkins and the Appalachian Trail Conservancy, *Along the Appalachian Trail, New Jersey, New York, and Connecticut*, Images of America Series (Charleston, SC: Arcadia Publishing, 2014), 54. Two photographs of the Lemon Squeezer are shown.

New York-New Jersey Trail Conference, *New York Walk Book*, 6th ed. (New York: New York-New Jersey Trail Conference, 1998), 282. The Lemon Squeezer is "a curious rock formation."

Irish Potato 127

Type of Formation: Large Boulder
WOW Factor: 8
Location: Willow Grove (Rockland County)
Tenth Edition, NYS Atlas & Gazetteer: p. 108, D1
Earlier Edition, NYS Atlas & Gazetteer: p. 32, BC3
Parking GPS Coordinates: 41°13.786'N, 74°03.621'W
Irish Potato GPS Coordinates: 41°14.151'N, 74°03.643'W
Accessibility: 0.6-mile hike
Degree of Difficulty: Moderate
Additional Information: Lake Welch Beach, 800 Kanawauke Road, Stony Point, NY 10980 (845) 947-2444
Trail Maps: Northern Harriman Bear Mountain Trails, Trail Map 4
Southern Harriman Bear Mountain Trails, Trail Map 3
parks.ny.gov/documents/parks/HarrimanTrailMap.pdf

Description

The Irish Potato is a massive, free-standing, 25-foot-high boulder whose shape and folds vaguely resemble that of a potato. The rock is located near the summit of 1,174-foot-high Irish Mountain.

History

The Irish Potato was given its name by Bill Hoeferlin, the man who started the Hikers Region Map series.

Directions

From the Palisades Interstate Parkway, take Exit 14 for Haverstraw and drive northwest on Route 98/Willow Grove Road for 1.5 miles. From the point where Route 106/Gate Hill Road enters on your right, continue

55. Irish Potato. Photograph by Dan Balogh.

northwest on Route 106/Gate Hill Road for another 0.2 mile and park in a pull-off on your left.

From the parking area, cross over the road and follow the yellow-blazed Suffern-Bear Mountain Trail north for 0.6 mile, initially paralleling Minisceongo Creek. The rock is at the junction with the informal path coming up from the Beaver Pond Campground at Lake Welch.

Resources

"Irish and Pound Swamp Mountains, Harriman State Park (5/1/2010)." http://www.danbalogh.com/irish.html. A series of photographs of the Irish Potato are shown.

"Harriman State Park, NY – Irish Potato and Hansenclever Winter Loop." https://gonehikin.blogspot.com/2018/12/harriman-state-park-ny-irish-potato-and.html.

Three Witches

128

Type of Formation: Large Boulder
WOW Factor: 4
Location: Willow Grove (Rockland County)
Tenth Edition, NYS Atlas & Gazetteer: p. 108, D1
Earlier Edition, NYS Atlas & Gazetteer: p. 32, BC3
Parking GPS Coordinates: 41°13.786′N, 74°03.621′W
Three Witches GPS Coordinates: 41°13.486′N, 74°03.862′W (Google Earth)
Accessibility: 0.7-mile-long hike
Degree of Difficulty: Moderate
Trail Maps: parks.ny.gov/documents/parks/HarrimanTrailMap.pdf
Southern Harriman Bear Mountain Trails, Trail Map 3

Description

The Three Witches (possibly named after the three Witches featured in Shakespeare's play *Macbeth*) consist of three, large glacial erratics somewhat aligned in a row. Some hikers refer to them as Moe, Larry, and Curly—a nod to the Three Stooges.

History

The Suffern-Bear Mountain Trail takes you through the former estate of George Briggs Buchanan, who was vice president of the Corn Products Refining Company, producer of Karo syrup. Buchanan named his estate Orak, which is Karo spelled backwards. Clever, huh?

Directions

From the Palisades Interstate Parkway, take Exit 14 for Haverstraw, and drive northwest on Route 98/Willow Grove Road for 1.5 miles. From where Route 106/Gate Hill Road enters on your right, continue northwest on

Route 106/Gate Hill Road/Kanawauke Road for another 0.2 mile and park in a pull-off on your left.

From the pull-off, scamper down the north bank of Minisceongo Creek and rock hop across the stream. Once on the south side of the creek, pick up the yellow-blazed Suffern-Bear Mountain Trail and head west uphill for 0.1 mile. When you come to an old, abandoned paved road, proceed south for 0.1 mile, following the yellow-blazed markers. At a junction, bear left onto an old, secondary road, continuing to follow the yellow-blazed trail/road southeast for another 0.4 mile, passing by the ruins of George Briggs Buchanan's estate along the way. The three, aligned glacial erratics are on the left side of the trail, <0.7 mile from the start of the hike.

Resources

New York-New Jersey Trail Conference, *New York Walk Book*, 6th ed. (New York: New York-New Jersey Trail Conference, 1998), 300. The book describes the Three Witches as "large glacial erratics."

Harriman State Park: Southern Section

There are a seemingly endless number of glacial erratics in Harriman State Park. The ones described here are those that have been given colorful names, have interesting histories, or have become notable destinations or way points along hikes. The rocks also tend to be trailside.

Claudius Smith's Cave 129

Type of Formation: Shelter Cave
WOW Factor: 7–8
Location: Tuxedo Park (Orange County)
Tenth Edition, NYS Atlas & Gazetteer: p. 107, D9–10
Earlier Edition, NYS Atlas & Gazetteer: p. 32, C2
Parking GPS Coordinates: 41°11.781′N, 74°11.043′W
Claudius Smith's Cave GPS Coordinates: 41°11.867′N, 74°10.078′W
Accessibility: 1.6-mile hike
Degree of Difficulty: Moderate
Trail Maps: Southern Harriman Bear Mountain Trails, Trail Map 3
parks.ny.gov/documents/parks/HarrimanTrailMap.pdf

56. Interior of Claudius Smith's Cave. Photograph by the author.

Description

The Claudius Smith's Cave, aka Claudius Smith's Den, is described by Patricia Edwards Clyne in *Caves for Kids in Historic New York*: "The upper cave is formed by a horizontal crack at the base of an imposing cliff. It is 30 feet long, 8 feet deep and 8 feet high at its tallest point. Part way across the entrance, Smith built a protective wall of rock, the remains of which can still be seen. From this cave a winding passageway, with an excellent observation post about a third of the way up, leads through the rocks to the top of the cliff."

The top of the cliff is commonly called Claudius Smith's Rock.

In *Rock Scenery of the Hudson Highlands and Palisades*, Jerome Wyckoff writes that the cave "formed as slabs weathered out of the cliff. The cave's size is made possible by wide joint spacing."

In *Suffern: 200 Years, 1773–1973*, additional descriptions of Claudius Smith's Cave are offered: "The lower cave had once been an Indian shelter and numerous prehistoric artifacts have been found there. The Smith gang used it to shelter their animals (the Horsestable). The upper cave was partially built up in front for protection. From the rear of the overhang there was a winding passageway which led to an escape route at the top of the hill."

Claudius Smith's Cave was just one of several caves that Smith used for hideouts. According to Patricia Edwards Clyne, in *Hudson Valley Trails and Tales*, "A few hundred feet downhill to the east, there is another cave called Horse Stable Rock Shelter, since it was used to harbor the gang's four-footed loot."

A Horse Stable Rock was also reputedly used by Smith and his gang, roughly 0.5 mile from Route 202 near Wesley.

Take note of a massive, 8-foot-high cube of rock that lies on the ground near the cave, evidently having broken off from the cliff face.

History

Claudius Smith was a real-life, Tory outlaw. He led a gang of lawless men, including three of his sons, William, Richard, and James, who raided settlers, retreating back to the cave to avoid capture. In the end, Smith's luck ran out. He was caught and hanged on January 22, 1779. Sons Richard and James met a similar fate at the gallows later while William died in a Goshen jail from wounds sustained in a shoot-out.

The cave is reputed to be the largest Indian rock-shelter in southern New York State. The Minsi Native Americans used the cave while on hunting expeditions. We know this because pottery shards and arrowheads have been found in the cave and nearby vicinity.

Directions

From south of Southfields (junction of Routes 17 & 17A), drive south on Route 17 for 2.5 miles. Turn left onto East Village Road, and then immediately left into a large parking area. From here, walk south along Route 17 to the Tuxedo Train Station. Follow the tracks south for a short distance, and then cross over the Ramapo River via a steel footbridge.

Once across the river, head north through a skate park. Then turn right onto East Village Road, go under I-87/NYS Thruway, and then turn immediately left onto Grove Drive, proceeding north for 0.1 mile to pick up the trailhead (41°11.699'N, 74°10.818'W) on your right.

Before starting your hike, take note that two paths heading in different directions begin from the trailhead. Ignore the white-blazed Kakiat Trail that goes off to your right.

Follow the red-blazed Ramapo-Dunderberg Trail east into the woods. At a junction, 1.2 miles from the Tuxedo Station, continue straight ahead on the red-blazed Tuxedo-Mt. Ivy Trail as the red-blazed Ramapo-Dunderberg Trail heads left. Pay attention at this junction, and make sure to stay on the trail leading off to your right. After another 0.4 mile, you will come to Claudius Smith's Cave, on your left. Both the upper and lower caves are worth a look.

The Blue Disk Trail, to your left, takes you to the summit above the caves, which some call Claudius Smith's Rock.

Resources

New York-New Jersey Trail Conference, *New York Walk Book*, 6th ed. (New York: New York-New Jersey Trail Conference, 1998). A line drawing of Claudius Smith's cave is on page 305. The cave is described as "an overhanging rock formation used as a hideout during the Revolutionary War."

Patricia Edwards Clyne, *Caves for Kids in Historic New York* (Monroe, NY: Library Research Associates, 1980), 63–75.

"Claudius Smith, 'Cowboy of the Ramapos,' Hangs." https://www.history.com/this-day-in-history/claudius-smith-cowboy-of-the-ramapos-hangs.

Jerome Wyckoff, *Rock Scenery of the Hudson Highlands and Palisades* (Glens Falls, NY: Adirondack Mountain Club, 1971). A photograph of the rock can be seen on pages 4 and 34.

New York-New Jersey Trail Conference, *Day Walker: 32 Hikes in the New York Metropolitan Area*, 2nd ed. (Mahwah, NJ: New York-New Jersey Trail Conference, 2002). A photograph of the cave is shown on page 176, followed by a chapter on Claudius Smith, his exploits, and the cave, pages 177–183.

Philip H. Smith, *Legends of the Shawangunk and its Environs* (Fleischmanns, NY: Purple Mountain Press, 1967), 60–65.

Village of Suffern Bicentennial Committee, *Suffern: 200 Years, 1773–1973* (Suffern, NY: Village of Suffern Bicentennial Committee, 1973), 80–84. "Bandits of the Ramapos" goes into the history of Smith and his gang of lawless men.

Patricia Edwards Clyne, *Hudson Valley Trails and Tales* (Woodstock, NY: Overlook Press, 1990), 174–181.

Clay Perry, *Underground Empire: Wonders and Tales of New York Caves* (New York: Stephen Daye Press, 1948), 137–138.

Raymond H. Torrey, Frank Place Jr., and Robert L. Dickinson, *New York Walk Book*, 3rd ed. (New York: American Geographical Society, 1951), 132–133. "Here are two dens of Claudius Smith and a number of smaller shelters at intervals along the summit. The former are horizontal cracks at the foot of successive cliffs. The upper are 8 to 10 feet high and 30 feet long with a depth of 10 feet. While the lower is longer and deeper but not so high."

Orange-Ulster Board of Cooperative Education Services, *Orange County: A Journey through Time* (Orange County, NY: reprint, 1984). A photograph of Claudius Smith's cave can be seen on page 64.

Cornelia F. Bedell, *Now and Then and Long Ago in Rockland County* (Rockland County, NY: Privately printed, 1941), 10.

Paul M. Ochojski, ed., *More Gleanings from Rockland History* (Orangeburg, NY: Journal of the Historical Society of Rockland County, 1971). "The Ballad of Claudius Smith" by H. Pierson Mapes is recounted on pages 10–12.

Bonora's Rock, Elbow Brush, and Large Unnamed Boulder

130

Type of Formation: Large Boulder; Narrow passageway
WOW Factor: 5
Location: Tuxedo Park (Orange County)
Tenth Edition, NYS Atlas & Gazetteer: p. 107, D9
Earlier Edition, NYS Atlas & Gazetteer: p. 32, C2–3
Parking GPS Coordinates: *Tuxedo Park*: 41°11.781′N, 74°11.043′W; *Johnsontown Road Circle*: 41°10.803′N, 74°09.822′W
Destination GPS Coordinates: *Bonora's Rock*: not determined; *Elbow Brush*: 41°11.653′N, 74°10.121′W (guesstimate); *Large, Unnamed Boulder*: not determined; *Almost Perpendicular*: 41°11.055′N, 74°10.276′W (guesstimate)
Accessibility: From Claudius Smith's Cave: *Elbow Brush*: 0.4-mile hike; *Bonora's Rock*: >0.4-mile hike; *Top of Almost Perpendicular*: 1.1-mile hike
From Johnsontown Circle: *Top of Almost Perpendicular*: 0.9-mile hike; *Bonora's Rock & Elbow Brush*: 1.6-mile hike; *Claudius Smith's Cave*: 2.0-mile hike
Degree of Difficulty: Moderately difficult
Trail Map: parks.ny.gov/documents/parks/HarrimanTrailMap.pdf

Description

According to William J. Myles, in *Harriman Trails: A Guide and History*, Bonora's Rock, aka Rockneath, is a very large rock. Unfortunately, we have no specific information as to its exact size.

Elbow Brush is a narrow passageway that is so constricted that hikers must take off their backpacks in order to squeeze through. A spur path conveniently bypasses Elbow Brush for those who are stouter or less adventurous.

57. Elbow Brush. Photograph by Dan Balogh.

Large, Unnamed Boulder: Although this large rock is associated with Native Americans, we have not been able to find out any specifics about it regarding its size.

History

In 1928, Bonora's Rock was named for John Bonora, who was a member of the Paterson Rambler's Club. Bonora loved the rock so much that he built a furnished cabin that lay against the massive rock. The cabin came to be known as Rockneath.

According to the Village of Suffern Bicentennial Committee, "the large boulder served as an Algonquin shelter."

Directions

JOHNSONTOWN CIRCLE APPROACH

From Sloatsburg (junction of Seven Lakes Drive and Route 17), drive northeast on Seven Lakes Drive for 0.5 mile. As soon as you pass under I-87/NYS Thruway, turn left onto Johnsontown Road and head northeast for 1.7 miles. Park at the cul-de-sac.

Follow the Blue Disc Trail. In 0.8 mile you will reach the top of Almost Perpendicular (a steep climb); at <1.2 mile, coming down from Pound Mountain cliff, look for a spur path that leads quickly to Bonora's Rock; at 1.4 miles, you will cross a gas pipeline corridor; at 1.7 miles, Elbow Brush is encountered; then, at 2.0 miles, you will come to Claudius Smith's Cave.

CLAUDIUS SMITH'S CAVE APPROACH

See chapter on Claudius Smith's Cave for directions to Claudius Smith's Cave. From the rocky top of Claudius Smith's Cave, follow the Blue Disc Trail south. After <0.4 mile, you will pass through a narrow passage called Elbow Brush. At 0.7 mile, a gas pipeline corridor is passed.

Head up Pound Mountain, following the Blue Disc Trail. At 0.2 mile before the mountain summit is reached, a short spur path leads to Bonora's Rock.

At 1.3 miles, you will come to the top of Almost Perpendicular and begin getting ready to descend. The name was given to this cliff by the Fresh Air Club.

The steep descent takes you down to the junction with the white-marked Kakiat Trail, "just below a huge boulder" at 1.6 miles. From here, stay on the Blue Disc Trail for another 0.5 mile to reach the Johnsontown Circle.

Resources

Village of Suffern Bicentennial Committee, *Suffern: 200 Years, 1773–1973* (Suffern, NY: Village of Suffern Bicentennial Committee, 1973), 84.

William J. Myles and Daniel Chazin, *Harriman Trails: A Guide and History*, 4th ed. (New York: New York-New Jersey Trail Conference, 2018). On page 27 is an account of Bonora's rustic cabin. On page 27 is also a photograph of hikers ascending the very steep, near-vertical slope of "Almost Perpendicular."

"Almost Perpendicular, Claudius Smith Den and Elbow Brush – Harriman State Park." https://takeahike.us/almost-perpendicular-claudius-smith-den-elbow-brush-harriman-state-park/.

"Harriman – Almost Perpendicular, Elbow Brush" https://www.njhiking.com/harriman-almost-perpendicular-elbow-brush/.

Large Boulder near Johnsontown Road Cul-De-Sac

131

Type of Formation: Large Boulder
WOW Factor: 5
Location: Sloatsburg (Orange County)
Tenth Edition, NYS Atlas & Gazetteer: p. 107, D9–10
Earlier Edition, NYS Atlas & Gazetteer: p. 32, C2
Parking GPS Coordinates: 41°10.803′N, 74°09.822′W
Large Boulder near Cul-de-sac GPS Coordinates: 41°10.784′N, 74°09.876′W
Accessibility: 0.05-mile walk
Degree of Difficulty: Easy
Trail Maps: Southern Harriman Bear Mountain Trails, Trail Map 3 parks.ny.gov/documents/parks/HarrimanTrailMap.pdf

---

Description

This large boulder is frequently seen by hikers at the start of the Blue Disc Trail.

Although the boulder could have been included with the rocks described in the previous chapter, we have given it its own chapter for hikers with mobility issues who want to see a large, woods-enclosed boulder near the roadside.

History

We have not been able to find out any specifics about the boulder's size or history.

What we do know is that at an earlier time, Johnsontown Road continued farther north beyond the cul-de-sac. This latter section was abandoned in 1962 when Seven Lakes Drive was constructed.

58. Large boulder near Blue Disc Trailhead. Photograph by Dan Balogh.

Directions

From Sloatsburg (junction of Seven Lakes Drive and Route 17), drive northeast on Seven Lakes Drive for 0.5 mile. As soon as you pass under I-87/NYS Thruway, turn left onto Johnsontown Road and head northeast for 1.7 miles. Park at the cul-de-sac.

From the cul-de-sac, walk back along the road for 100 feet, crossing over a tiny stream, and turn right at the start of the Blue-Disc Trail. The large rock is immediately to your left.

Horsestable Rock 132

WOW Factor: 4
Location: Montebello (Rockland County)
Tenth Edition, NYS Atlas & Gazetteer: p. 107, E10
Earlier Edition, NYS Atlas & Gazetteer: p. 32, CD2–3
Parking GPS Coordinates: 41°08.748′N, 74°06.760′W
Horsestable Rock GPS Coordinates: 41°09.828′N, 74°06.127′W (Google Earth)
Accessibility: 2.2-mile hike
Degree of Difficulty: Moderate
Trail Maps: Southern Harriman Bear Mountain Trails, Trail Map 3
parks.ny.gov/documents/parks/HarrimanTrailMap.pdf
rocklandgov.com/files/7413/4555/9420/Parks_Kakiat_Map.pdf

Description

Horsestable Rock is a large boulder that sits close to the gas pipeline corridor that runs northwest/southeast in the woods above Montebello. Cornelia F. Bedell, in *Now and Then and Long Ago in Rockland County*, points out that Horse Stable Rock, of Revolutionary War fame, "weigh[s] many hundreds of tons."

The rock was named for its proximity to Horsestable Mountain.

History

The 376-acre Kakiat County Park was created in 1972, eleven years after the county board of supervisors acquired 239 acres of land from Anthony Cuccolo who owned Kakiat Farm.

"Kakiat" is Native American for *kackyachtaweke*, a word that means a "draw"—a place where one side of a stream is open, and the other side bounded by a hill.

Directions

From Suffern (junction of Routes 202/Wayne Avenue & 59/Orange Avenue), take Route 202/Wayne Avenue northeast for 3.0 miles. Opposite the Viola Elementary School, turn left into the Kakiat County Park and park in the designated parking area.

From the parking area, cross over the bridge spanning the Mahwah River. Follow the white-blazed Kakiat Trail northwest for 0.9 mile. When you come to the gas pipeline corridor, turn right, and follow the corridor northeast for 1.2 miles. Horsestable Rock will be on your right, less than 35 feet from the edge of the woods.

Resources

Cornelia F. Bedell, *Now and Then and Long Ago in Rockland County, New York* (New City, NY: Historical Society of Rockland County, 1968), 10.

South of the Mountain: The Historical Society of Rockland County 27, no. 2 (April–June, 1983). A photograph of the rock along with Julian Harris Salomon standing by it can be seen on page 21.

The Egg

133

Type of Formation: Large Rock

WOW Factor: 5

Location: Montebello (Orange County)

Tenth Edition, NYS Atlas & Gazetteer: p. 107, D10

Earlier Edition, NYS Atlas & Gazetteer: p. 32, CD2–3

Parking GPS Coordinates: 41°08.748′N, 74°06.760′W

Destination GPS Coordinates: *Rock at junction of Suffern-Bear Mountain Trail and Conklins Crossing:* 41°10.162′N, 74°06.879′W (guesstimate); *The Egg*: 41°10.137′N, 74°06.616′W; *Boulder near the Egg*: 41°10.132′N, 74°06.630′W

Accessibility: 2.8-mile hike

Degree of Difficulty: Moderate

Trail Maps: Southern Harriman Bear Mountain Trails, Trail Map 3 parks.ny.gov/documents/parks/HarrimanTrailMap.pdf

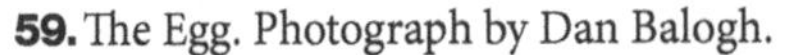

59. The Egg. Photograph by Dan Balogh.

Description

The Egg is a massive dome of rock jutting out from the woods. Some writers have described it as a "great boulder," but we think of it as a massive, rounded mound of exposed bedrock.

Along the way to the Egg, two large boulders are encountered at the junction of the Suffern-Bear Mountain Trail and Conklins Crossing (named for Ramsey Conklin).

A boulder of some size is also encountered 80 feet southwest of the Egg.

Directions

From Suffern (junction of Routes 202/Wayne Avenue and 59/Orange Avenue), take Route 202/Wayne Avenue/Haverstraw Road northeast for 3.0 miles. Opposite the Viola Elementary School, turn left into the Kakiat County Park and park in the designated parking area.

From the parking area, cross over the bridge spanning the Mahwah River and follow the white-blazed Kakiat Trail northwest for 1.5 miles. At a junction, turn right onto the yellow-blazed Suffern-Bear Mountain Trail and proceed northeast. After another 1.1 miles, you will pass by the Conklins Crossing junction. In another 0.2 mile, the Egg is reached, 2.8 miles from the trailhead.

The Egg can also be reached starting from the Suffern-Bear Mountain Trail trailhead, but it involves a trek of 5.8 miles.

Resources

"New York, NY: Southern Harriman Loop." https://www.backpacker.com/trips/new-york-ny-southern-harriman-loop.

Stone Giant 134

Type of Formation: Large Boulder; Rock-Shelter
WOW Factor: 6
Location: Sloatsburg (Orange County)
Tenth Edition, NYS Atlas & Gazetteer: p. 107, D10
Earlier Edition, NYS Atlas & Gazetteer: p. 32, CD2–3
Parking GPS Coordinates: 41°10.430′N, 74°10.118′W
Destination GPS Coordinates: *Glacial Erratics*: Not determined; *Stone Giant #1*: 41°10.495′N, 74°08.263′W; *Stone Giant #2*: 41°10.513′N, 74°08.253′W (Google Earth)
Accessibility: *Glacial Erratics*: 0.4-mile hike; *Stone Giant:* 1.8-mile hike
Degree of Difficulty: Moderate
Trail Maps: Southern Harriman Bear Mountain Trails, Trail Map 3
parks.ny.gov/documents/parks/HarrimanTrailMap.pdf

Description

According to Stella Green and H. Neil Zimmerman, in *50 Hikes in the Lower Hudson Valley*, "Two huge erratics frame a small waterfall on the brook."

The Stone Giant, aka Ga-Nus-Quah Rock, is a grouping of large, 15-foot-high rocks next to Pine Meadow Brook. Google Earth shows two boulders that are relatively close together.

In *Suffern: 200 Years, 1773–1973*, the Stone Giant site is described as "an Indian rockshelter with a huge boulder known as 'Ganusquah (Stone Giant).'" In addition, "there is a fireplace here, and rocky cliffs to examine."

An old photograph of the Stone Giant shows the profile of a Native American face with a broad, protruding nose, which is perhaps the feature that gave the rock its anthropomorphic name.

Directions

From north of Sloatsburg (junction of Seven Lakes Drive & Route17/Orange Avenue), drive northeast on Seven Lakes Drive for 1.6 miles and turn right into the Reeves Meadow Information Center, named for Robert Reeves, who once operated a farm in the area.

Starting from the parking area, follow the red-blazed Pine Meadow Trail east. Just before you reach the point at 0.4 mile where the yellow-blazed Stony Brook Path comes in on your left, take note of two large erratics that Stella Green and H. Neil Zimmerman refer to in *50 Hikes in the Lower Hudson Valley*.

At 0.6 mile, the trail crosses the Columbia Gas Transmission Company pipeline; at 1.2 miles, the Hillburn-Torne-Sebago Trail is reached and passed, as well as a footbridge at 1.6 miles. Finally, at 1.8 miles, you will come to two large rocks—the Stone Giants—which are on the right side of the trail. Also look for the rock-shelter that is reported in *Suffern: 200 Years, 1773–1973*.

Resources

New York-New Jersey Trail Conference, *Day Walker: 32 Hikes in the New York Metropolitan Area*, 2nd ed. (Mahwah, NJ: New York-New Jersey Trail Conference, 2002), 173.

Village of Suffern Bicentennial Committee, *Suffern: 200 Years, 1773–1973* (Suffern, NY: Village of Suffern Bicentennial Committee, 1973), 77. A photograph of the Stone Giant can be seen on page 79.

William J. Myles, *Harriman Trails: A Guide and History* (New York: The New York-New Jersey Trail Conference, 1994). A photograph of William Myles taken at Ga-Nus-Quah Rock in 1936 is shown on page 91.

Stella Green and H. Neil Zimmerman, *50 Hikes in the Lower Hudson Valley* (Woodstock, VT: Backcountry Guides, 2002), 116.

"Ga-Nus Quah Rocks Bouldering." https://www.mountainproject.com/area/107403866/ga-nus-quah-rocks.

Leonard M. Adkins and the Appalachian Trail Conservancy, *Along the Appalachian Trail, New Jersey, New York, and Connecticut*, Images of America Series (Charleston, SC: Arcadia Publishing, 2014), 51. A hiker is shown leaning against the rock.

New York-New Jersey Trail Conference, *New York Walk Book*, 6th ed. (New York: New York-New Jersey Trail Conference, 1998), 292.

The Rockland Record, vol. 2 (Rockland County, NY: Rockland County Society of the State of New York, 1931). On page 93 is a photograph of what appears to be the Stone Giant with the caption "Colonel Rockledge Stone, an old 'Residenter' of the Ramapo Mountains." Perhaps this is yet another name for the Stone Giant.

Raccoon Brook Hill Caves 135

Type of Formation: Talus Cave
WOW Factor: 5
Location: Sloatsburg (Rockland County)
Tenth Edition, NYS Atlas & Gazetteer: p. 107, D10
Earlier Edition, NYS Atlas & Gazetteer: p. 32, C2–3
Parking GPS Coordinates: 41°10.430′N, 74°10.118′W
Raccoon Brook Hill Caves GPS Coordinates: 41°10.376′N, 74°08.243′W (Google Earth)
Accessibility: 2.0-mile hike
Degree of Difficulty: Moderately difficult
Trail Maps: Southern Harriman Bear Mountain Trails, Trail Map 3 parks.ny.gov/documents/parks/HarrimanTrailMap.pdf

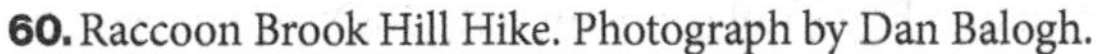

60. Raccoon Brook Hill Hike. Photograph by Dan Balogh.

Description

In *Suffern: 200 Years, 1773–1973*, the Village of Suffern Bicentennial Committee describe Raccoon Brook Hill Caves as a "jumble of rock boulders" that contain a number of tiny caves.

History

According to William J. Myles and Daniel Chazin, in *Harriman Trails: A Guide and History*, Native American artifacts have been found at the site and are now on exhibit at the Bear Mountain Trailside Museum.

Directions

From north of Sloatsburg (junction of Seven Lakes Drive & Route 17/ Orange Avenue), drive northeast on Seven Lakes Drive for 1.6 miles and turn right into the Reeves Meadow Information Center.

From the information center, follow the red-blazed Pine Meadows Trail east for 1.6 miles, along the way passing by a gas pipeline corridor at 0.6 mile, and the Hillburn-Torre-Sebago Trail at 1.2 miles. Be sure to stay on the south side of Pine Meadow Brook. At a junction with the Kakiat Trail, bear right and follow the Kakiat Trail southeast for 0.3 mile. Then turn left onto the black-marked Raccoon Brook Hills Trail. You will come to the caves within 0.05 mile, literally at the foot of Raccoon Hill. If you arrive at the yellow-blazed Poached Egg Trail (on your left), then you have gone almost 0.5 mile too far along the Raccoon Brook Hills Trail.

Resources

Village of Suffern Bicentennial Committee, *Suffern: 200 Years, 1773–1973* (Suffern, NY: Village of Suffern Bicentennial Committee, 1973), 78.

"Raccoon Brook Hill Hike." https://nycdayhiking.com/hikes/raccoonbrook.htm.

William J. Myles and Daniel Chazin, *Harriman Trails: A Guide and History*, 4th ed. (New York: New York-New Jersey Trail Conference, 2018), 88.

The Pulpit 136

Type of Formation: Large Rock
WOW Factor: Unknown
Location: Tuxedo (Orange County)
Tenth Edition, NYS Atlas & Gazetteer: p. 107, D10
Earlier Edition, NYS Atlas & Gazetteer: p. 32, C2–3
Parking GPS Coordinates: 41°10.430'N, 74°10.118'W
The Pulpit GPS Coordinates: Not determined
Accessibility: >3.0-mile hike
Degree of Difficulty: Moderately difficult
Trail Maps: Southern Harriman Bear Mountain Trails, Trail Map parks.ny.gov/documents/parks/HarrimanTrailMap.pdf

---

Description

The Pulpit is a large rock that juts out from the cliff edge on top of a bluff.

History

The Raccoon Brook Trail was blazed in 1931 by Paul Scubert. He named it "Trail of the Raccoon Hills." According to William J. Myles, in *Harriman Trails: A Guide and History*, the jutting rock was called the Pulpit by early hikers, but no credit is given to any one hiker for naming the rock.

Directions

From north of Sloatsburg (junction of Seven Lakes Drive & Route 17/ Orange Avenue), drive northeast on Seven Lakes Drive for 1.6 miles and turn right into the Reeves Meadow Information Center.

From the information center parking area, walk 0.2 mile southwest on the Pine Meadow Trail. Then turn left onto the blue-blazed Seven Hills Trail and head first south, and then north, for 2.6 miles.

When you come to the junction with the Raccoon Brook Hills Trail, turn right and head east on the Raccoon Hill Brook Trail. At 0.3 mile you will pass by the Reeves Brook Trail, which enters on your left. Soon after, you will cross over a small brook and then begin climbing a steep fault scarp. Look up to see the Pulpit, high above you, 0.5 mile from the start of the Raccoon Hill Brook Trail.

Resources

William J. Myles, *Harriman Trails: A Guide and History* (New York: New York-New Jersey Trail Conference, 1994), 99–100.

Monitor Rock and Cracked Diamond

137

Type of Formation: Large Boulder
WOW Factor: 7
Location: Sloatsburg (Rockland County)
Tenth Edition, NYS Atlas & Gazetteer: p. 107, D10
Earlier Edition, NYS Atlas & Gazetteer: p. 32, C2–3
Parking GPS Coordinates: 41°11.920′N, 74°07.773′W
Destination GPS Coordinates: *Monitor Rock*: 41°11.401′N, 74°07.489′W; *Cracked Diamond*: 41°11.324′N, 74°07.457′W (estimated); *Two other boulders*: 41°11.368′N, 74°07.499′W
Accessibility: *Monitor Rock*: 0.7-mile hike; *Cracked Diamond*: 0.8-mile hike
Degree of Difficulty: Moderately easy
Trail Maps: Southern Harriman Bear Mountain Trails, Trail Map 3
parks.ny.gov/documents/parks/HarrimanTrailMap.pdf

Description

In *Harriman Trails: A Guide and History*, William J. Myles describes Monitor Rock as a "large quartz boulder on a ledge." Pictures of it show a huge overhang, and a tight squeeze between two ledges.

Raymond H. Torrey, Frank Place Jr., and Robert L. Dickinson, in *New York Walk Book*, draw attention to "a ledge surmounted by a six-foot white quartz boulder."

Cracked Diamond is a very large, tilted rock that William Myles calls a "great boulder." The rock is located on a ridge along Diamond Mountain, directly to the left of the trail.

A number of other decent-sized erratics are located in the general area.

61. Monitor Rock Hike. Photograph by Dan Balogh.

History

Monitor Rock was named by the Fresh Air Club after a copy of the *Christian Science Monitor* was found tucked away under the rock.

During the days when Woodtown Road was drivable, the Seven Hills Trail originally started from Monitor Rock. The road is now a horse trail.

Diamond Mountain was called Halfway Mountain until 1927.

Directions

From Sloatsburg (junction of Seven Lakes Drive & Route 17/Orange Avenue), drive northeast on Seven Lakes Drive for 4.3 miles. Turn left into a parking area for Lake Sebago (a man-made lake created in 1925 by the damming up of Stony Brook and the Emmetfield Swamp).

Walk across Seven Lakes Drive and head south on the blue-blazed Seven Hills Trail for 0.6 mile. Turn left (west) onto the former Woodtown Road and quickly cross Diamond Creek. After 0.1 mile, the trail leaves the old road, and then turns sharply right.

Follow a white-blazed spur path to the left that leads to Monitor Rock on a ledge in 300 feet.

Continue following the Seven Hills Trail for another 0.1 mile. You will come to the Cracked Diamond Rock 0.1 mile before the intersection with the Tuxedo-Mt. Ivy Trail.

According to Charlie Stein, in his photoblog lowmileage.com/2013/06/30/harriman-state-park-rockland-county-new-york-2, a number of large boulders can be seen along this hike, of which Cracked Diamond Rock is just one of many.

Resources

William J. Myles, *Harriman Trails: A Guide and History* (New York: New York-New Jersey Trail Conference, 1994), 126.

"Diamond Mountain Loop from Lake Sebago Boat Launch – Harriman State Park." https://scenesfromthetrail.com/2017/08/15/diamond-mountain-loop-from-lake-sebago-boat-launch-harriman-state-park. This site contains excellent photographs of Monitor Rock and Cracked Diamond.

"Harriman State Park, Rockland County, New York." https://lowmileage.com/2013/06/30/harriman-state-park-rockland-county-new-york-2.

Raymond H. Torrey, Frank Place Jr., and Robert L. Dickinson, *New York Walk Book*, 3rd ed. (New York: American Geographical Society, 1951), 177.

High Hill Trail Boulder 138

Type of Formation: Large Boulder
WOW Factor: 4
Location: Sloatsburg (Rockland County)
Tenth Edition, NYS Atlas & Gazetteer: p. 107, D10
Earlier Edition, NYS Atlas & Gazetteer: p. 32, C2–3
Parking GPS Coordinates: 41°12.637′N, 74°07.289′W
High Hill Trail Boulder GPS Coordinates: 41°12.588′N, 74°07.284′W
Accessibility: Roadside or near roadside
Degree of Difficulty: Easy

Description

The only references to the High Hill Trail Boulder that we can find are taken from Raymond H. Torrey, Frank Place Jr., and Robert L. Dickinson's book, *New York Walk Book* and William J. Myles and Daniel Chazin's *Harriman Trails: A Guide and History*. Both call it, simply, a "great rock." Apparently, a cairn on top of the boulder, frequently obscured by bushes, announces the beginning of the High Hill Trail that leads towards the summit of 1,200-foot-high High Hill from the boulder.

Directions

From north of Sloatsburg (junction of Routes 17 & 72), drive north on Route 17 for <1.0 mile. At a traffic intersection, turn right onto Seven Lakes Drive and head northeast for <5.5 miles. Just before the point where two lanes of the highway divide, park to the right side of the road.

The large boulder is 250 feet south of the road division, on the east bank (right side) of the road.

The High Hill Trail leading from the boulder still exists, but has grown faint, so you may need to take a few minutes to find it.

NOTE: If roadside parking is an issue here, a second option is to turn left onto the Baker Camp Access Road marked by a sign reading "Camp Service Road," 0.5 mile south from where the Seven Lakes Drive divides, and head north for 0.3 mile. Park off-road to your right where the road bends sharply left and walk along an old dirt road northeast for <0.2 mile. Bear right, and scramble down to Seven Lakes Drive, just below you. This maneuver places you in the approximate location of the large boulder and the beginning of the slowly vanishing High Hill Trail.

Resources

Raymond H. Torrey, Frank Place Jr., and Robert L. Dickinson, *New York Walk Book*, 3rd ed. (New York: American Geographical Society, 1951), 288, 419.

William J. Myles and Daniel Chazin, *Harriman Trails: A Guide and History*, 4th ed. (New York: New York-New Jersey Trail Conference, 2018), 247.

Glacial Boulder, Kitchen Stairs, Valley of Dry Bones, The Odd Couple, MacIlvain's Rocks, Grandpa and Grandma Rocks, Lean-on-me, and the Egg

139

Type of Formation: Large Boulder

WOW Factor: 4–7

Location: Suffern (Rockland County)

Tenth Edition, NYS Atlas & Gazetteer: p. 107, E10

Earlier Edition, NYS Atlas & Gazetteer: p. 32, CD2–3

Parking GPS Coordinates: 41°07.017′N, 74°09.315′W

Trailhead GPS Coordinates: 41°07.134′N, 74°09.465′W

Destination GPS Coordinates: *Glacial Boulder*: 41°07.548′N, 74°09.157′W (Google Earth); *Kitchen Stairs*: 41°07.912′N, 74°08.891′W (Google Earth); *Valley of Dry Bones*: 41°08.963′N, 74°08.022′W (Google Earth); *The Odd Couple*: 41°08.986′N, 74°08.006′W (Google Earth); *MacIlvain's Rocks* and *Grandpa and Grandma Rocks*: 41°09.558′N, 74°07.304′W (Google Earth); *Lean-on-me*: 41°09.558′N, 74°07.304′W (Google Earth); *The Egg*: 41°10.148′N, 74°006.638′W (Google Earth)

Accessibility: *Glacial Boulder*: 1.0-mile hike; *Kitchen Stairs*: 1.5-mile hike; *Valley of Dry Bones*: 3.2-mile hike; *The Odd Couple:* 3.3-mile hike; *MacIlvain's Rocks*: 4.3-mile hike; *Grandpa and Grandpa Rocks*: 4.4-mile hike; *Lean-on-me*: 4.8-mile hike; *The Egg*: 5.8-mile hike

Degree of Difficulty: Moderate to difficult

Trail Maps: Southern Harriman Bear Mountain Trails, Trail Map 3 parks.ny.gov/documents/parks/HarrimanTrailMap.pdf

62. Huge rock by Kitchen Stairs. Photograph by Dan Balogh.

Description

A photograph of this unnamed boulder (Glacial Boulder) reveals a standalone, inclined rock some 6–8 feet high.

Kitchen Stairs is described as "a broken fault face," which we assume to be a modest fracture in the Earth's crust.

Valley of Dry Bones is a field of (relatively speaking) small boulders. It was named for the way in which the rocks resemble a graveyard of animal bones—a reference, apparently, to Ezekiel, chapter 37, in the Bible.

The Odd Couple are two large boulders in close proximity to one another, but unevenly matched size-wise.

MacIlvain's Rocks are a field of smaller rocks named after William J. MacIlvain of the Fresh Air Club, who managed to lead his group through what seemed to everyone to be, in retrospect, the most difficult route possible.

Grandpa and Grandma Rocks are two large boulders, one particularly pointed, surrounded by a retinue of grandchildren (smaller boulders). They were named by Frank Place and Raymond Torrey in 1926 while scouting the Suffern-Bear Mountain Trail.

Lean-on-me is a large, oval shaped boulder on a platform of exposed bedrock. Its front end rests on a smaller boulder. It is this feature, it would seem, that gives the rock its name.

The Egg is called "a great boulder" by William J. Myles in *Harriman Trails: A Guide and History*, undoubtedly because of its size and oval shape. To us, it looks like a large dome of exposed bedrock.

History

Kitchen Stairs and Valley of Dry Bones were named by Frank Place in 1925.

Directions

At Suffern (junction of Routes 59/Orange Avenue & 202/Wayne Avenue), park in the commuter parking area off Route 59 just south of the Thruway

63. Grandpa Rock. Photograph by Dan Balogh.

overpass. If hikers are still only allowed to park there on weekends, then you will need to find roadside parking on streets in Suffern if you are hiking on a weekday.

Begin walking north on Route 59/Orange Avenue/Orange Turnpike. As soon as you have passed under I-87/New York State Thruway, continue north for <0.1 mile and pick up the Suffern-Bear Mountain trailhead, on your right, where a square-shaped wooden sign indicates trail mileages.

Follow the Suffern-Bear Mountain Trail (SBM) northeast. At 1.4 miles, you will ascend a rock formation known as Kitchen Stairs. At 3.2 miles, you will come to the Valley of Dry Bones, where many boulders litter the woods along the trail.

In <0.05 mile further you will reach the Odd Couple, two large glacial boulders of different sizes in close proximity to one another.

After 4.4 miles (or 0.2 mile from crossing the Columbia Gas Company pipeline corridor and passing through the MacIlvain's Rocks), you will come to the Grandpa and Grandma Rocks.

Within another 0.4 mile, Lean-on-me Rock is encountered.

At 5.8 miles, the Egg is reached soon after you pass by the white-blazed Conklin Crossing, entering on your left. The Egg, if it is all that it is cracked up to be, provides a high point with views.

Resources

New York-New Jersey Trail Conference, *New York Walk Book*, 6th ed. (New York: New York-New Jersey Trail Conference, 1998), 299. Mention is made of "an area of giant boulders" soon followed by "two big boulders, called Grandpa and Grandma Rocks."

William J. Myles and Daniel Chazin, *Harriman Trails: A Guide and History*, 4th ed. (New York: New York-New Jersey Trail Conference, 2018). MacIlvain's Rocks, 118–119; Kitchen Stairs, 48, 118–119; Valley of Dry Bones, 118–119; the Egg, 35, 119.

"Harriman State Park, Rockland County, New York." https://lowmileage.com/2014/09/01/harriman-state-park-rockland-county-new-york-3. The site contains photographs of some unnamed rocks as well as the Valley of Dry Bones.

Raymond H. Torrey, Frank Place Jr., and Robert L. Dickinson, *New York Walk Book*, 3rd ed. (New York: American Geographical Society, 1951), 179.

West Pointing Rock 140

Type of Formation: Large Boulder
WOW Factor: 5
Location: Ladentown (Rockland County)
Tenth Edition, NYS Atlas & Gazetteer: p. 108, D1
Earlier Edition, NYS Atlas & Gazetteer: p. 32, C3
Parking GPS Coordinates: 41°11.112′N, 74°04.478′W
West Pointing Rock GPS Coordinates: Not determined
Accessibility: 3.0-mile hike
Degree of Difficulty: Moderately difficult
Trail Maps: Southern Harriman Bear Mountain Trails, Trail Map 3
parks.ny.gov/documents/parks/HarrimanTrailMap.pdf

Description

William J. Myles and Daniel Chazin, in *Harriman Trails: A Guide and History*, describes the rock as "a 10 x 14-foot boulder, with a sharp projection on its west side."

Directions

From the Palisades Interstate Parkway, take Exit 13 for Route 202/Haverstraw/Suffern, and drive west on Route 202 for 1.6 miles. Turn right onto Ladentown Road and head north for 0.2 mile. Then bear left onto Mountain Road and proceed northwest for 0.2 mile. When you come to Diltzes Road, turn left and drive southwest for 0.2 mile. Finally, bear right into a parking area 100 feet before coming to a power station.

From the parking area, follow the red-marked Tuxedo-Mt. Ivy Trail northwest for 3.0 miles. Bear right onto the white-blazed Breakneck Mountain Trail and head northeast for 0.2 mile to reach West Pointing Rock.

Stella Green and H. Neil Zimmerman in *50 Hikes in the Lower Hudson Valley* describe "an open slab with large boulders" along the trail, which may be West Pointing Rock or another rock formation entirely.

Resources

William J. Myles and Daniel Chazin, *Harriman Trails: A Guide and History*, 4th ed. (New York: New York-New Jersey Trail Conference, 2018). A photograph is shown on page 30.

Stella Green and H. Neil Zimmerman, *50 Hikes in the Lower Hudson Valley* (Woodstock, VT: Backcountry Guides, 2002), 119.

"New York-New Jersey Trail Conference." https://www.nynjtc.org/hike/print/6045.

64. West Pointing Rock. Photograph by Dan Balogh.

Iron Mountain Erratic 141

Type of Formation: Large Boulder
WOW Factor: 6
Location: Ladentown (Rockland County)
Tenth Edition, NYS Atlas & Gazetteer: p. 108, D1
Earlier Edition, NYS Atlas & Gazetteer: p. 32, C3
Parking GPS Coordinates: 41°11.112'N, 74°04.478'W
Iron Mountain Erratic GPS Coordinates: Not determined
Accessibility: <2.0-mile hike
Degree of Difficulty: Moderately difficult

Description

The quirky Iron Mountain Erratic is a 10-foot-high boulder standing upright. If you look at the photograph accompanying this chapter and use a little imagination, the rock seemingly turns into a human foot complete with toes.

Directions

From the Palisades Interstate Parkway, take Exit 13 for Route 202/ Haverstraw/Suffern, and drive west on Route 202 for 1.6 miles. Turn right onto Ladentown Road and head north for 0.2 mile. Then turn left onto Mountain Road and proceed northwest for 0.2 mile. When you come to Diltzes Road, turn left and drive southwest for 0.2 mile. Finally, bear right into a parking area 100 feet before coming to a power station.

From the parking area, follow the red-marked Tuxedo-Mt. Ivy Trail northeast along a gas pipeline corridor for 0.2 mile. When the Tuxedo-Mt. Ivy Trail veers off to the left, continue following the gas pipeline for another 0.7 mile, or roughly 1.0 mile from the parking area. The unmarked Iron Mountain Trail begins here, on the left, as an old woods road.

When you reach the top of Iron Mountain, look for the glacial erratic along the ridge line that extends north/south.

Resources

William J. Myles and Daniel Chazin, *Harriman Trails: A Guide and History*, 4th ed. (New York: New York-New Jersey Trail Conference, 2018. On page 250, they describe the rock as a "large near vertical bolder."

65. Iron Mountain Erratic. Photograph by Dan Balogh.

Valley of Boulders, Great Boulders, Indian Rock-Shelter, and Bowling Rocks

142

Type of Formation: Large Rock; Rock-Shelter
WOW Factor: 4–5
Location: Southfields (Orange County)
Tenth Edition, NYS Atlas & Gazetteer: p. 107, D9
Earlier Edition, NYS Atlas & Gazetteer: p. 32, BC2
Parking GPS Coordinates: 41°14.507′N, 74°10.534′W
Destination GPS Coordinates: *Valley of Boulders*: 41°14.991′N, 74°09.060′W (estimated); *Great Boulders*: Not determined; *Indian Rock-Shelter*: 41°14.844′N, 74°08.988′W (estimated); *Bowling Rocks*: 41°14.647′N, 74°07.572′W (Google Earth)
Accessibility: *Valley of Boulders*: 1.0-mile hike; *Great Boulders*: 1.9-mile hike; *Indian Rock-Shelter*: 2.2-mile hike; *Bowling Rocks*: 4.2-mile hike
Degree of Difficulty: Moderately difficult
Trail Maps: Southern Harriman Bear Mountain Trails, Trail Map 3
parks.ny.gov/documents/parks/HarrimanTrailMap.pdf

Description

Valley of Boulders consists of a number of large rocks that have calved off from a low-lying escarpment ridge. William J. Myles, in *Harriman Trails: A Guide and History* writes, "The ravine of the brook is known as Valley of Boulders."

Great Boulders, size-wise, are encountered along the left side of the trail.

Indian Rock-Shelter is an impressive rock-shelter that overlooks Green Pond, providing a nifty place to stop and have a bite or two, or to just take a breather. According to Stella Green and H. Neil Zimmerman, in *50 Hikes in the Lower Hudson Valley*, Green Pond is slowly atrophying and it is only a matter of time until the pond becomes Green Swamp.

Per William J. Myles, "The boulders that dot the bare rock give rise to the name 'Bowling Rocks.'"

History

The Nurian Trail was named for Kerson Nurian, a Bulgarian electrical engineer who worked on submarines in the Brooklyn Navy Yard. He constructed the trail around 1929.

Directions

From Southfields (junction of Routes 17 & 19/Orange Turnpike), drive south on Route 17 for <0.2 mile and turn left into the parking area for the Southfields Post Office, making sure to park as unobtrusively as possible. Next to the post office is Spring Street. Proceeding now on foot, turn right onto Spring Street, walk north for >0.1 mile, and then right onto Railroad Avenue, which takes you to the end of the road and railroad tracks within several hundred feet.

Follow the railroad tracks north for >0.1 mile to reach the Nurian Trailhead [41°14.766'N, 74°10.365'W]. Look for triple blazes on a post. Follow the trail as it veers right, quickly crossing over the Ramapo River via a footbridge and then over the New York State Thruway via the Southfields Pedestrian Bridge, aka Nurian Bridge. By now, you will have gone roughly 0.4 mile from the parking area. Head east on the white-blazed Nurian Trail, crossing over a paved road at 1.4 miles and then, further on, trekking through the Valley of Boulders (a ravine). Great Boulders are passed at 1.9 miles.

INDIAN ROCK SHELTER

At 2.0 miles, turn right onto the yellow-marked Dunning Trail (named for James M. Dunning, past chairman of the AMC Trails Committee), which quickly takes you to the Indian Rock-Shelter in less than 0.2 mile.

BOWLING ROCKS

Continue southeast on the yellow-blazed Dunning Trail from the second junction with the Nurian Trail. At 1.1 miles from the start of the Dunning Trail, the White Bar Trail junction is reached. Bear left onto the White

Bar Trail, continuing also on the Dunning Trail. Shortly after, the two separate. Stay right on the Dunning Trail. At 1.7 miles, the Dunning Trail is crossed by the red-blazed Ramapo-Dunderberg Trail. Just a short way beyond are the Bowling Rocks, on your left.

Resources

William J. Myles, *Harriman Trails: A Guide and History* (New York: New York-New Jersey Trail Conference, 1994), 41, 85.

Stella Green and H. Neil Zimmerman, *50 Hikes in the Lower Hudson Valley* (Woodstock, VT: Backcountry Guides, 2002), 122.

"Hiking Island Pond and Valley of Boulders Loop." https://www.stavislost.com/hikes/trail/harriman-state-park-island-pond-and-valley-of-boulders-loop.

"Map Index – Harriman-Bear Mountain Trails Map Set." https://www.nynjtc.org/content/index-harriman-bear-mountain-trails-map. Contains index for maps.

Bergen County

Bergen County occupies the northeastern corner of the state of New Jersey and borders the Hudson River across from Manhattan, the Bronx, and the southern part of Westchester County. It encompasses 671 square miles, only 5.5 percent being water.

Lamont Rock 143

Rockleigh Woods Sanctuary and Lamont Reserve

Type of Formation: Large Boulder
WOW Factor: 6
Location: South of Orangeburg (Bergen County, NJ)
Tenth Edition, NYS Atlas & Gazetteer: p. 110, A5
Earlier Edition, NYS Atlas & Gazetteer: p. 32, D4
Parking GPS Coordinates: 41°00.241'N, 73°55.532'W
Destination GPS Coordinates: *Lamont Rock*: Not determined; *Large Rock*: 40°59.683'N, 73°55.269'W; *Rocky area*: 40°58.669'N, 73°55.761'W
Accessibility: >1.0-mile hike
Degree of Difficulty: Moderate
Additional Information: Rockleigh Municipal Building, 26 Rockleigh Road, Rockleigh, NJ 07647
Trail Map: nynjtc.org/sites/default/files/RockleighLamontTrailMap_Color2015.pdf

---

Description

The Lamont Rock is a 10-foot-high boulder with a pronounced lean to it. From certain angles, it looks like the rock has nosed-dived right into the ground.

History

Both the Rockleigh Woods Sanctuary and Lamont Reserve were once part of Camp Alpine of the Greater New York Councils, Boy Scouts of America. The 84-acre Rockleigh Woods Sanctuary was purchased in 1975, the 134-acre Lamont Reserve in 1996.

66. Lamont Rock. Photograph by Dan Balogh.

Directions

From the Palisades Interstate Parkway, take Exit 4 for Route 9W and proceed north on Route 9W/Palisades Boulevard for 1.2 miles. Turn left onto Oak Tree Road and head west for >0.1 mile. Then turn left onto Closter Road, which, after crossing under the Palisades Interstate Parkway and then crossing the NY/NJ border, becomes Rockleigh Road at 0.6 mile.

In another <0.3 mile, turn left into the driveway for the Rockleigh Municipal Building opposite Willow Avenue and park at the rear of the building.

From the rear of the playground area next to where you parked, follow the blue-blazed Hutcheon Trail into the woods. At 0.05 mile, bear left at a fork and follow the yellow/blue-blazed Sneden-Harding-Lamont Trail east for 0.1 mile. When you come to the next fork, stay left as the blue-blazed Hutcheon Trail veers off to the right, and follow the yellow-blazed Sneden-Harding-Lamont Tail as it now proceeds southeast. In another 0.6 mile, you will pass by the red-blazed Roaring Brook Trail to your right and then cross over Roaring Brook.

Immediately after the stream crossing, you will pass by the white-blazed Lamont Rock Trail on your right. In <0.05 mile, turn left onto the white-blazed Lamont Rock Trail as it departs from the yellow-blazed Sneden-Harding-Lamont Trail, and follow it to the southeast corner of the preserve. From here, round the corner and continue west on the white-blazed Lamont Rock Trail for another 0.2 mile until you come to Lamont Rock.

Resources

"Rockleigh Woods Sanctuary and Lamont Reserve Loop." https://www.nynjtc.org/hike/rockleigh-woods-sanctuary-and-lamont-reserve-loop.

Haring Rock

144

Tenafly Nature Center/Lost Brook Preserve

Type of Formation: Large Rock
WOW Factor: 6
Location: Tenafly (Bergen County, NJ)
Tenth Edition, NYS Atlas & Gazetteer: p. 110, BC5
Earlier Edition, NYS Atlas & Gazetteer: p. 24, AB4
Parking GPS Coordinates: *East Clinton Avenue*: 40°54.476′N, 73°56.649′W;
Tenafly Nature Center: 40°55.455′N, 73°56.677′W
Haring Rock GPS Coordinates: 40°54.500′N, 73°56.608′W (estimated)
Accessibility: 200-foot walk from East Clinton Avenue
Degree of Difficulty: Easy
Trail Map: tenaflynaturecenter.org/Trail-Map
Trail maps can also be obtained at the visitor center at 313 Hudson Avenue, Tenafly, NJ 07670

Description

John Serrao, in *The Wild Palisades of the Hudson*, writes, "'Haring Rock,' a huge sandstone erratic which is 10 feet tall, 10 feet around, and weighs 15 tons, is visible on a marked trail off East Clinton Avenue." It is said to be the largest sandstone glacial erratic in the region.

A second rock formation, called Little Rock Den, can also be seen along the Allison Trail, 500 feet before the junction with the orange-blazed Haring Rock Trail.

There seem to be a number of sizable boulders in the preserve. Within a 150-foot radius to the east of Haring Rock are a number of large rocks.

67. Haring Rock. Photograph by Dan Balogh.

History

Haring Rock is named for John J. Haring, an early twentieth-century physician, who would on occasion stop at the rock to take a break while on his journeys.

Haring Rock has not been at its present location for very long. Community developers relocated the boulder in order to make room for the Jewish Community Center, now 0.3 mile south, after which the rock was cemented in place upside down to ensure that it would stay put.

Both the Haring Rock Trail and the Seely Trail come together at Haring Rock.

The Lost Preserve, where the rock is located, and the Tenafly Nature Center Preserve, were founded in 1961. The Lost Preserve encompasses 330 acres of land, the Tenafly Nature Center Preserve 52 acres.

Directions

EAST CLINTON AVENUE

From southeast of Tenafly (junction of Routes 9W and East Clinton Avenue), drive northwest on East Clinton Avenue for 0.5 mile and park in a small pull-off on your right.

From the pull-off, follow the orange-marked Haring Rock Trail northeast for 250 feet. When you come to the junction with the yellow/orange-marked Seely Trail, the rock will be on your right.

TENAFLY NATURE CENTER

From southeast of Tenafly (junction of Routes 9W and East Clinton Avenue), drive northwest on East Clinton Avenue for 1.8 miles. Turn right onto Engle Street and head northeast for <1.8 miles. When you come to Hudson Street, bear right and proceed southeast for 0.7 mile to reach the Tenafly Nature Center.

Resources

John Serrao, *The Wild Palisades of the Hudson* (Westwood, NJ: Lind Publications, 1949), 10.

"Tenafly Nature Center Loop #2 – New York – New Jersey." https://www.nynjtc.org/hike/tenafly-nature-center-loop-2.

"Moving Haring Rock, 1979." https://tpl.omeka.net/items/show/181. This website offers a series of photographs that show how the rock was relocated.

The Palisades

68. Palisades. Antique postcard, public domain.

The following section goes into detail about a number of unusual rocks to be found along the Palisades—a line of steep, high cliffs along the west side of the Hudson River that provide dramatic views of the Manhattan skyline. The Palisades, aka New Jersey Palisades and Hudson River Palisades, extend from Jersey City, New Jersey, north to near Nyack, New York, a distance of about twenty miles.

In *The Hudson: From the Wilderness to the Sea*, Benson Lossing writes, "Between Piermont and Hoboken, these rocks present, for a considerable distance, an uninterrupted, rude, columnar front, from appearance, yet not actually so in form. They have a steep slope of debris, which has been crumbling from the cliffs above, during long centuries, by the action of frost and the elements. The ridge is narrow, being in some places not more than three-fourths of a mile in width. It is really an enormous projecting trap-dyke."

In the 1882 book, *History of Bergen and Passaic Counties, New Jersey*, mention is made that "the height of the range near Weehawken is about three hundred and ten feet above the river, rising gradually to five hundred and forty feet near the northern terminus."

The Palisades Interstate Park was created in 1900 to help preserve and safeguard the Palisades for future generations to enjoy. Until then, companies had been free to quarry the rock, and the Palisades were ultimately in danger of being obliterated. At one time, there were seventeen quarries operating between Fort Lee and the state line.

Protecting the Palisades proved to be a prodigious job, for most of the Palisade lands were in private ownership, with large summer estates and woodlands overlooking the top of the Palisades, and several riverfront villages below. As in so many cases involving eminent domain, sacrifices had to be made by individual landowners.

The word palisades comes from the Latin word palus, or "stake." With a little imagination, the Palisades can be seen as a wooden palisaded fortification. The Dutch called the Palisades the Dutch equivalent of "Great Chip Rocks." The Lenape, a Native American tribe, called the Palisades Wee-Awh-En, meaning "rocks that look like trees."

The rock that forms the Palisades is diabase, more commonly known as traprock (which comes from the Swedish word trapp for "stairs"). Similar basalt columns can be seen at the Devil's Tower in Wyoming; East Rock in New Haven, Connecticut; and the Giant's Causeway in Northern Ireland.

The hamlet of Palisades was earlier known as Snedens Landing.

Resources

Benson Lossing, The Hudson: From the Wilderness to the Sea (Sommersworth, NY: New Hampshire Publishing Company, 1972; facsimile of the 1866 edition), 360.

Jeffrey Perls, Paths along the Hudson: A Guide to Walking and Biking (New Brunswick, NJ: Rutgers University Press, 2001), 152–153.

Station Rock 145

Type of Formation: Large Rock
WOW Factor: 2–3
Location: Palisades (Rockland County)
Tenth Edition, NYS Atlas & Gazetteer: p. 111, B5–6
Earlier Edition, NYS Atlas & Gazetteer: p. 25, A4
Parking GPS Coordinates: 40°59.327'N, 73°54.421'W
Border between Rockland County and Bergen County, NJ: 40°59.840'N, 073°54.163'W
Station Rock GPS Coordinates: 40°59.984'N, 73°54.162'W (estimated)
Accessibility: 1.8-mile hike (estimated)
Degree of Difficulty: Moderately difficult
Palisades Interstate Park Map: njpalisades.org/pdfs/bywayMap.pdf.

Description

According to Alice Munro Haagensen, in *Palisades and Snedens Landing*, "Marking the boundary [between New York and New Jersey] was a great rock near the water in the middle of which was chiseled a line and the words 'Latitude 41°North.' On the south side were marked the words 'New Jersey' and on the north, 'New York.'" Haagensen continues by declaring that "considering the many graffiti on nearby rocks, it seems wise to keep its whereabouts a little vague." It is for this reason, undoubtedly, that an incomplete GPS reading of 40°59'51.20" is given in Haagensen's book.

History

Station Rock is significant, for it marks the early attempts of surveyors to delineate the boundary line between New York and New Jersey at the Palisades. As it turned out, the true boundary ended up being about 900 feet farther south than the 41st parallel. However, the boundary line still passes through Station Rock, giving New York an extra 10 square miles of land.

One source contends that a pole was set into place next to the rock in 1930. It is no longer present.

We see that there is also a Station Rock at the top of the Palisade cliffs, roughly 50 feet from the main trail. According to Arthur G. Adams, in *The Hudson River Guidebook*, "Here, atop the cliffs, is a 6-foot memorial shaft erected in 1882."

Directions

Heading north on the Palisades Interstate Parkway, 1.7 miles beyond exit 2, turn right after seeing a sign for the State Line Lookout.

Heading south on the Palisades Interstate Parkway, go 0.4 mile past exit 3 and then turn left onto a U-turn that takes you onto the northbound lane of the parkway. Go 0.4 mile and get off at the sign for the State Line Lookout.

From either approach, drive northeast for 0.6 mile on the road that leads to the parking area for the State Line Lookout.

From east of the State Line Café, walk north along an old cement road (now a pedestrian walkway) for <0.2 mile. Bear right onto the aqua-blazed Long Path and follow it downhill for 0.8 mile, descending to a lower section of the Palisades.

At a junction indicated by a sign that states "U.S. Route 9W/Lamont 0.3 mile. Peanut Leap Cascade 0.3 mile. The Giant Stairs 1.1 miles," bear right, and follow the white-blazed Shore Trail downhill for >0.3 mile to the bottom of the Palisades and base of Peanut Leap Falls. Peanut Leap Falls is an 80-foot-high, seasonal cascade that flows down a near-vertical wall of chiseled, black, palisaded rock. It is here that you will also see the ruins of the Lawrence-Tonetti Gardens that were constructed in the nineteenth century by Mary Lawrence and her husband, François Tonetti.

Turn right and continue south on the Shore Trail for <0.5 mile. You will know that you are in the general area of Station Rock when you come to a chain-link fence that once delineated the boundary between Bergen County (NJ) and Rockland County (NY).

Station Rock is several hundred feet before you reach the chain-link fence.

Resources

Alice Munro Haagensen, *Palisades and Snedens Landing* (Tarrytown, NY: Pilgrimage Publishing, 1986), 26.

Wikipedia, s.v. "New York – New Jersey Line War," last modified September 28, 2022, https://en.wikipedia.org/wiki/New_York_-_New_Jersey_Line_War. This website details the New York-New Jersey line war that endured for 64 years.

Arthur G. Adams, *The Hudson River Guidebook* (NY: Fordham University Press, 1996), 124.

Giant Stairs 146

Type of Formation: Huge Talus Slope
WOW Factor: 9
Location: Palisades (Rockland County)
Tenth Edition, NYS Atlas & Gazetteer: p. 111, A6
Earlier Edition, NYS Atlas & Gazetteer: p. 25, A4–5
Parking GPS Coordinates: 40°59.327′N, 73°54.421′W
Giant Stairs GPS Coordinates: 40°59.086′N, 73°54.344′W
Accessibility: 4.3-mile round trip
Degree of Difficulty: Difficult
Trail Map: njpalisades.org/pdfs/map.pdf

69. Giant Stairs. Photograph by Dan Balogh.

Description

Giant Stairs, aka "Giant's Stairs" and "Stairway to the Sun," is a >0.6-mile-long section of the Palisades Shore Trail characterized by small to enormous boulders that must be scrambled over and around. According to Arthur G. Adams, in *The Hudson River Guidebook*, the Giant's Stairway is a "natural stone stairway up cliffs, used as a hiking route."

You will see many boulders and pieces of talus. Take note that many of these pieces of rock have been there for thousands of years. The ones lacking vegetation are of more recent origin.

History

The Palisades Shore Trail, which the Giant Stairs is on, has been a favorite hike for many decades. At 13.5 miles in length, it is also the single longest trail along the Hudson River's shoreline. The trail was designated a National Recreation Trail in 1971.

It should be noted that the Palisades, due to geological forces, are endlessly reshaping themselves. It was as recently as 2012 that 10,000 tons of rock broke off from the cliffs just south of the state line, leaving behind a 520-foot scar on the cliffs.

Directions

Follow the directions given in the chapter on Station Rock to reach the base of the Palisades at Peanut Leap Falls.

From Peanut Leap Falls, head southwest on the white-blazed Shore Trail for >0.9 mile to reach the beginning of the Giant Stairs.

After 0.6 mile of scrambling and rock-hopping, you will reach the south end of the talus field/Giant Stairs. From here, continue south on the Shore Tail for another 0.3 mile. Then turn right onto the blue & white-blazed Forest View Trail and climb uphill, heading northwest for >0.3 mile. When you come to the aqua-blazed Long Path, bear right and proceed northeast for 0.6 mile to return to the parking area.

Resources

Herb Chong, ed., *The Long Path Guide*, 5th ed. (Mahwah, NJ: New York-New Jersey Trail Conference, 2002), 32. Trail information is provided.

Cy A. Adler, *Walking the Hudson, Batt to Bear: From the Battery to Bear Mountain* (Cathedral, NY: Green Eagle Press, 1997), 72.

"Giant Stairs – Hudson River Palisades – Scenes from the Trail." https://scenesfromthetrail.com/2017/07/19/giant-stairs-palisades-interstate-park.
"Giant Stairs/Long Path Loop from State Line." https://nynjtc.org/hike/giant-stairslong-path-loop-state-line-lookout.
New York-New Jersey Trail Conference, *Day Walker: 32 Hikes in the New York Metropolitan Area*, 2nd ed. (Mahwah, NJ: New York-New Jersey Trail Conference, 2002), 242.
"Giant Stairs Hike – Hudson River Palisades – Scenes from the Trail." https://scenesfromthetrail.com/2016/07/08/giant-stairs-hike-palisades-interstate-park.
"Palisades Interstate Park: Giant Stairs Hike." https://www.nycdayhiking.com/hikes/palisad3.htm.
Arthur G. Adams, *The Hudson River Guidebook* (NY: Fordham University Press, 1996), 124.

“Man-in-the-Rock” Pillar

147

Type of Formation: Pillar
WOW Factor: 5
Location: Palisades (Bergen County, NJ)
Tenth Edition, NYS Atlas & Gazetteer: p. 111, B6
Earlier Edition, NYS Atlas & Gazetteer: p. 25, A4–5
Parking GPS Coordinates: 40°59.327′N, 73°54.421′W
“Man-in-the-Rock” Pillar GPS Coordinates: 40°58.084′N, 73°54.700′W (estimated)
Accessibility: 1.5-mile hike
Degree of Difficulty: Moderate
Additional Information: Map titled “Hiking at Alpine Picnic Area (North)” shows the location of the Man in the Rock: njpalisades.org/pdfs/hikeAlpineNorth.pdf. A second map is at njpalisades.org/pdfs/hikeStateline.pdf.

Description

The Man-in-the-Rock Pillar is a towering rock column that leans against the side of the rock wall. In *The Hudson River Guidebook*, Arthur G. Adams describes it as “the highest, most isolated and conspicuous pillar of rock in the Palisades, literally curving 70 feet high between two large slides.” This same, nearly identical description is found in Raymond H. Torrey, Frank Place Jr., and Robert L. Dickinson’s *New York Walk Book*.

History

The distinctive Man-in-the-Rock Pillar is located at Bombay Hook, aka Boompes Hook and Bumpy Hook—a prominent bend in the Hudson River north of Alpine.

The pillar’s name comes from the face of a man that supposedly can be seen near the base of the pillar’s north side.

Directions

See chapter on Station Rock for directions to the State Line Lookout parking area.

From the parking area, follow the aqua-blazed Long Path south for 1.5 mile to reach the pillar.

Raymond H. Torrey, Frank Place Jr., and Robert L. Dickinson mention that it is far easier to see the full scope of this formation from a boat rather than when you are standing next to it, close-up.

Resources

"Rock Climb Pillar, Palisades Park." https://www.mountainproject.com/route/106098610/pillar.

Arthur G. Adams, *The Hudson River Guidebook* (NY: Fordham University Press, 1996), 124.

"Palisades Bombay Hook." https://www.geocaching.com/geocache/GC2D748_palisades-bombay-hook.

New York-New Jersey Trail Conference, *New York Walk Book*, 6th ed. (New York: New York-New Jersey Trail Conference, 1998).

Raymond H. Torrey, Frank Place Jr., and Robert L. Dickinson, *New York Walk Book*, 3rd ed. (New York: American Geographical Society, 1951). A pen-and-ink drawing done by Robert L. Dickinson, entitled "Cleft above Bombay Hook," is displayed on page 30. On page 36, the writers indicate that the Man-in-the-Rock is the northern column of Bombay Hook.

"Hiking at Alpine Picnic Area (North)." https://www.njpalisades.org/pdfs/hikeAlpineNorth.pdf.

Gray Crag 148

Type of Formation: Rock Pillar
WOW Factor: 5
Location: Alpine (Bergen County, NJ)
Tenth Edition, NYS Atlas & Gazetteer: p. 110, B5
Earlier Edition, NYS Atlas & Gazetteer: p. 25, A4
Parking GPS Coordinates: 40°57.200′N, 73°55.237′W
End of Ruckman Road GPS Coordinates: 40°58.470′N, 73°54.652′W
Gray Crag GPS Coordinates: 40°58.557′N, 73°54.605′W (estimated)
Accessibility: 1.9-mile hike
Degree of Difficulty: Moderate.

Description

According to Jeffrey Perls in *Paths along the Hudson: A Guide to Walking and Biking*, "Gray Crag is the largest isolated section of the Palisades cliff. It is more than three-hundred feet long and can be reached via a bridge that crosses the narrow ravine that separates it from the main cliff."

The concrete bridge, supported by a pair of steel I-beams, is about 30 feet long and spans a ravine on the west side of the detached rock pillar that is over 20–30 feet deep.

Although the rock pillar is long, it is fairly narrow, no more than 10–20 feet in width. What is dramatic is its height. It stands about 40 stories above the river on its east side!

History

In 1918, John Ringling (of Ringling Brothers fame) purchased two properties to create a 100-acre estate that he and his wife, Mable, named Gray Crag. On it they built an elegant, 20-room summer home which they used through the 1920s. It was demolished in 1935. Ruins of it are still visible today.

Directions

Driving north on the Palisades Interstate Parkway, take Exit 2 and head south for 0.2 mile to reach the park headquarters parking area, on your left.

Driving south on the Palisades Interstate Parkway, take Exit 2. Turn right onto Route 9 and head north for 0.1 mile. At the first light, turn right onto Alpine Approach Road and head southeast for >0.1 mile, going underneath two overpass bridges. At a fork, turn right and drive south for 0.1 mile to reach the park headquarters, on your left.

From the Palisades Interstate Park Commission's main office in New Jersey, follow the Long Path north for 1.7 miles until you come to the end of old Ruckman Road. The following directions are borrowed from Herb Chong's *The Long Path Guide*: "The Long Path turns left on Ruckman Road and, in another 50 feet, turns right on a gravel road into well-developed forest. Meet a second gravel road leading right to run along the cliff edge. This road, not part of the Long Path, ends in about 900 feet at the terminus of a great split off the main face of the Palisades." You are at Gray Crag.

Resources

Jeffrey Perls, *Paths along the Hudson: A Guide to Walking and Biking* (New Brunswick, NJ: Rutgers University Press, 2001), 163.

"Gray Crag. Palisades Interstate Park in New Jersey." https://www.njpalisades.org/graycrag.html.

"Palisades: John Ringling's 'Gray Crag.'" https://www.geocaching.com/geocache/GC2D746_palisades-john-ringlings-grey-crag.

Herb Chong, ed., *The Long Path Guide*, 5th ed. (Mahwah, NJ: New York-New Jersey Trail Conference, 2002), 22.

Raymond H. Torrey, Frank Place Jr., and Robert L. Dickinson, *New York Walk Book*, 3rd ed. (New York: American Geographical Society, 1951), 37. "Up above, there stands the largest separated section of rock in the Palisades—Gray Crag, some 300 feet long and 10 to 20 feet wide." The writers go on to mention that directly north is Bombay Hook and the Man-in-the-Rock pillar.

"Gray Crag – Sites – eMuseum." https://emuseum.ringling.org/emuseum/sites/144/gray-crag. This website contains a photograph of John and Mable Ringling's summer home, Gray Crag.

Alpine Rock and Hay-Kee-Pook Rock

149

Type of Formation: Large Rock
WOW Factor: 3–4
Location: Alpine (Bergen County, NJ)
Tenth Edition, NYS Atlas & Gazetteer: p. 111, B6
Earlier Edition, NYS Atlas & Gazetteer: p. 25, A4–5
Parking GPS Coordinates: 40°56.678′N, 73°55.145′W
Destination GPS Coordinates: *Alpine Rock*: 40°56.799′N, 73°55.129′W;
Hay-Kee-Pock Rock: Not determined
Accessibility: 0.1-mile walk
Degree of Difficulty: Easy
Additional Information: Palisades Interstate Park Commission / New Jersey Offices, Alpine, NJ 07620; (201) 768-1360
Alpine Picnic Area & Boat Basin Map: njpalisades.org/pdfs/hikeAlpine-South.pdf

---

Description

Of Alpine Rock, Cy A. Adler, in *Walking the Hudson, Batt to Bear: From the Battery to Bear Mountain*, writes, "Just south and in front of the old house [Kearney House] with its peeling white paint is a large, cubical rock. This black chunk of basalt, about 8 feet high, fell off the Palisades cliffs in 1896 and rolled onto the attached kitchen which it completely demolished. Assuming the rock is thrice the density of water, this slight chip off the Palisades weighs over 70,000 pounds."

It is rare that you get to see a rock that, after breaking off from a cliff, does such damage to a downhill structure. The Alpine Rock is roughly 50 feet southeast of the Kearney House, partially concealed by brush or a tree.

70. Rocks along Alpine section of Shore Trail. Photograph by Dan Balogh.

Hay-Kee-Pook Rock is also in the general area. In their *New York Walk Book*, Raymond H. Torrey, Frank Place Jr., and Robert L. Dickinson write, "South of the Landing is a path full of variety and charm, which winds up and down owing to washouts on the river edge way. The big boulder by the river is known as 'Hay-Kee-Pook,' and legend has it that an Indian lover committed suicide" by the rock.

History

The Kearney House, aka Blackledge-Kearney House, and Lord Charles Cornwallis's Headquarters, is very historic. The southern part of the house was constructed in the 1760s, the northern addition about 1840.

The house was first owned by Maria Blackledge and her husband, Daniel Van Sciver, and then later by James and Rachel Kearney, who ran it as a tavern. It was from these former occupants that the house came to have two names.

The house was purchased by the Palisades Interstate Park in 1907 and was added to the National Register of Historic Places in 1984.

There is a story that Lord Cornwallis used the house as his temporary headquarters while pursuing the Continental Army in 1776, although not everyone is convinced of the story's veracity.

Directions

Follow the directions given in the chapter on Gray Crag to reach the park headquarters [40°57.172'N, 73°55.219'W].

From the park headquarters, drive south down Henry Hudson Drive for 0.9 mile. When you come to a traffic circle, follow the sign pointing the way to the Alpine Picnic Area and Boat Basin and proceed north, continuing downhill on Henry Hudson Drive for another 0.3 mile to reach the parking area.

ALPINE ROCK

From the north end of the Alpine parking area, walk north for 0.1 mile. The Kearney House is to your left, opposite the second, smaller boat basin.

The boulder is near the south side of the house, no more than 50 feet away.

HAY-KEE-POOK ROCK

The Hay-Kee-Pook Rock is said to be south of Alpine Landing and the Boat Basin.

Resources

Cy A. Adler, *Walking the Hudson, Batt to Bear: From the Battery to Bear Mountain* (Cathedral, NY: Green Eagle Press, 1997), 68.

"Revolutionary War Sites in Alpine, New Jersey." https://www.revolutionary-warnewjersey.com/new_jersey_revolutionary_war_sites/towns/alpine_nj_revolutionary_war_sites.htm.

Jeffrey Perls, *Paths along the Hudson: A Guide to Walking and Biking* (New Brunswick, NJ: Rutgers University Press, 2001), 159.

"Crag Now & Then." https://njpalisades.org/pdfs/kearneyBrochure.pdf.

Raymond H. Torrey, Frank Place Jr., and Robert L. Dickinson, *New York Walk Book*, 3rd ed. (New York: American Geographical Society, 1951), 37.

Sampson's Rock (Historic) 150

Type of Formation: Large Boulder
WOW Factor: 7
Location: Englewood Cliffs (Bergen County, NJ)
Tenth Edition, NYS Atlas & Gazetteer: p. 110, C5
Earlier Edition, NYS Atlas & Gazetteer: p. 24, AB4
Sampson's Rock GPS Coordinates: 40°53.064'N, 73°57.037'W
Accessibility: Roadside

Description

According to Raymond H. Torrey, Frank Place Jr., and Robert L. Dickinson, in *New York Walk Book*, "This perched boulder, which has been carried upward at least 160 feet from the meadows by the ice sheet, measures 8 by 12 by 12 feet and is of soft red sandstone touching a few points on the hard gray trap rock beneath." One source estimates the boulder's weight at 32 tons.

Arthur C. Mack, in *The Palisades of the Hudson*, writes, "The drift [of ice] resulted also in many rocky curiosities, conspicuous among them as an isolated block of Triassic sandstone, called Sampson's Rock, perched upon the flat trap directly east of Englewood. This huge boulder, measuring nearly twelve feet in diameter and weighing many tons, was lifted 160 feet up the western slope of the ridge by the ice (it is calculated), and finally dropped in to its present resting place. Its under surface still retains the polish it received through the attrition of that movement."

History

We are not able to find any current history on this glacial erratic, despite Torrey, Place and Dickinson's assertion that it is the "best erratic in New Jersey."

A plaque was once on the boulder, but then vandalized and finally taken and not replaced.

As a point of interest, there is a Sampson's Rock at Madison Center, Connecticut, at a GPS of 41°16.660'N, 72°35.738'W.

Directions

From the Palisades Interstate Parkway, take Exit 1 and drive west on Palisades Avenue for either 0.1 mile or 0.3 mile (depending upon which direction you were heading when you came off from the Palisades Interstate Parkway).

You will come to the junction of Palisades Avenue and Floyd Street.

Torrey, Place, and Dickinson write, "A pleasant detour can be made by turning west on Palisades Avenue one block to Floyd Street and then right past seven houses to the best erratic in New Jersey—Sampson's Rock standing in a vacant lot on the right."

These directions may have been accurate at the time the book was published, but times have changed over the last 70 years. The boulder is actually near the junction of Floyd Street and Demarest Avenue, on the right-hand side of Floyd Avenue, between two houses just before Demarest Avenue.

Take note that the giant rock is obviously on private land. Be respectful and do your looking from the road.

Resources

Raymond H. Torrey, Frank Place Jr., and Robert L. Dickinson, *New York Walk Book*, 3rd ed. (New York: American Geographical Society, 1951), 23. "The best of these 'erratics' is Sampson's Rock in Englewood Cliffs." Details about the rock are on page 36.

Washington's Head and Indian Head (Historic) 151

Type of Formation: Rock Profile
WOW Factor: 2–3
Location: Fort Lee (Bergen County, NJ)
Tenth Edition, NYS Atlas & Gazetteer: p. 110, C5
Earlier Edition, NYS Atlas & Gazetteer: p. 24, B4 or p. 24, A1–2
Parking GPS Coordinates: 40°51.639′N, 73°57.326′W
Washington Head & Indian Head GPS Coordinates: Unknown

71. Indian Head and the Palisades. Antique postcard, public domain.

Description

Two rock profiles once existed on the Palisades bluffs until they were blasted apart and destroyed by quarrying. Washington's Head was demolished around 1897, Indian Head in 1898.

In the case of Indian Head, we know exactly how the rock profile met its demise. A 5-foot-wide hole was bored 100 feet into the rock face and packed with 7,000 pounds of dynamite. The resulting explosion completely destroyed the rock formation, sending down 350,000 tons of diabase to be used for construction.

In *History of Bergen County, New Jersey*, J. M. Van Valen writes, "Indian Head, one of the most historic points of the Palisades, a few years ago projected one hundred and fifty feet into the North River [Hudson River] beyond the point." The destruction of Indian Head "was one of the most successful efforts ever made to destroy the grandeur of this part of the Hudson. It broke out an area surface of one hundred and seventy-five feet by one hundred and sixty-five-feet and a depth of about one hundred feet, constituting nearly one-third of the height of the cliff."

We assume that Washington's Head met a similar fate.

Even if these rock formations had survived nineteenth-century quarrying, most likely they would now be mere shadows of their former selves, eroded by over a century of wind, rain, snow, sleet, and ice.

A third rock face also existed along the Palisades, but only fleetingly. In *New York's Palisades Interstate Park*, Barbara H. Gottlock and Wesley Gottlock write about a rock slide around 1941 that created the image of Hitler's face. Fittingly, right after the end of World War II, another rockslide permanently erased the image.

It's hard to know what to make of synchronistic events like this.

According to Raymond H. Torrey, Frank Place Jr., and Robert L. Dickinson, in *New York Walk Book*, "North of Ross Dock and above the Henry Hudson Drive are several striking rock formations. An immense squat column stands on a base that gives a thrilling but false impression that it might crumble at any time." The writers also mention a "leaning column nearby dubbed 'Fallen Caesar,'" but we can find no modern references to this rock formation. The writers also call attention to a second Indian Head: "Looking south from Point Lookout along the ridge, one sights the face of Indian Head—hook-nosed, low forehead and, for feathers, brushes bent backwards." The formation is said to be between Forest View and State Line at the highest point in the Palisades.

History

Fort Lee Park encompasses 34 acres of land next to the Hudson River. It is the site of a former Revolutionary War fort.

It was at Fort Lee that Thomas Paine wrote one of his political papers, beginning with the words, "These are the times that try men's souls."

Directions

From the town of Fort Lee (junction of Routes 12/Main Street & 67/ Schlosser Street), head east on Route 12/Main Street for 0.3 mile, near the end, heading downhill and veering right. You will come out onto a continuation of Main Street/River Road. Head south for 0.2 mile. Then bear left onto Henry Hudson Drive and proceed north for 0.9 mile. At a traffic circle, stay right and continue north for another 0.3 mile to the parking area for the Ross Dock Picnic Park.

The rock formations are said to have been 1.0 mile north of Fort Lee, with Indian Head about 500 yards from Washington Point. This would place the former rock profiles in the general area of the Ross Dock Picnic Park.

If you wish to look for some of the pillar formations mentioned in the *New York Walk Book*, we would recommend walking north along the Shore Trail and looking up occasionally to your left.

Resources

C. R. Roseberry, *From Niagara to Montauk: The Scenic Pleasures of New York State* (Albany, NY: State University of New York Press, 1982), 254.

Barbara H. Gottlock and Wesley Gottlock, *New York's Palisades Interstate Park*, Images of America Series (Charleston, SC: Arcadia Publishing, 2007), 14.

"Shaped by Nature and Man: The Geological History of the Palisades." https://www.amnh.org/learn-teach/curriculum-collections/young-naturalist-awards/winning-essays/1998-2003/shaped-by-nature-and-man-the-geological-history-of-the-palisades.

J. M. Van Valen, *History of Bergen County, New Jersey* (New York: New Jersey Publishing and Engraving, 1900), 653.

Stanley Wilcox and H. W. Van Loan, *The Hudson from Troy to the Battery* (Philmont, NY: Riverview Publishing, 2011), 126.

Jeffrey Perls, *Paths along the Hudson: A Guide to Walking and Biking* (New Brunswick, NJ: Rutgers University Press, 2001), 155.

Raymond H. Torrey, Frank Place Jr., and Robert L. Dickinson, *New York Walk Book*, 3rd ed. (New York: American Geographical Society, 1951), 34, 39.

Arthur C. Mack, *The Palisades of the Hudson* (Edgewater, NJ: Palisade Press, 1909), 39. "The greater part of old Indian Head was blown asunder to be metamorphosed into flats and skyscrapers."
"Washington's Head Gone." *Morning Union*, May 13, 1893, https://cdnc.ucr.edu/?a=d&d=MU18930513.2.32.

Miscellaneous Rocks 152

1. Lewis Beach, *Cornwall* (Newburgh, NY: E. M. Ruttenber & Sons, 1873).

A number of boulders and unusual rocks are mentioned in Beach's book. Several of them are along West Point Road in the Giant's Haunt and in Idlewild.

p. 90. Chapel Rock: Beach mentions a stream where "its passage through the glen is checked at intervals by huge boulders: one of which is known as Chapel Rock."

p. 108. Lover's Rocking Stone: "This stone, weighing several tons, is so equipoised that a child of ten can move it from side to side. It is sufficiently low to the ground to form a convenient seat for lovers."

p. 109. Giant's Slipper: As the story is told, a mythical, humungous giant, "whilst descending . . . dropped his slipper, which, turned to stone over six feet in length, in perfect shape." It "is still to be seen and known as the Giant's Slipper."

p. 110. Picnic Rock: This rock is "a table summit of platform rock covered with moss and lichen, and provided with numerous blocks of stone of varying size, which answers for chairs and tables."

p. 115. Poised Rock "is a parallelogram in shape, having two sides, the upper and lower, ten feet in length, and the ends about four feet. It measures eighteen feet in circumference, and being formed of granite, will weigh about fourteen tons. This huge rock stands by itself alone, lifted entirely from the pedestal rock on which it rests, except at one point. This point of contact is not over four inches square. It is supported on the extreme westerly end by a flat stone, eight inches high—eighteen inches long and fifteen wide."

p. 172. Natural Bridge: This rock phenomenon spans a stream that flows into the southwest end of Poplopen's Pond. "Its breadth across the stream is fifty feet and its length up and down the stream about eighty feet. In times of drought people can pass under it."

2. Jerome Wyckoff, *Rock Scenery of the Hudson Highlands and Palisades* (Glens Falls, NY: Adirondack Mountain Club, 1971).

p. 31. Hogencamp Mine Boulder: This boulder consists of a block of rock that lies along the Dunning Trail near Hogencamp Mine (41°14.606′N, 74°07.167′W). A photograph of the rock, along with three hikers next to it, can be seen.

p. 33. Pine Hill Boulder: Large blocks of talus along a steep slope on Pine Hill are shown in a photograph. Also evident are a number of rocks along a dirt road paralleling a power line.

p. 84. Circle Mountain Boulder: A photograph of a rock on Circle Mountain shows a large boulder that resembles the North Salem Balance Rock.

3. Richard M. Lederer Jr., *The Place-Names of Westchester County, New York* (Harrison, NY: Harbor Hill Books, 1978).

p. 40. Devil's Den and Buzzard's Cave: These two formations are by a sheer cliff on the southeast side of Byram Mountain, south of Byram Lake Road. If it's of any help, the GPS reading for nearby Byram Lake Reservoir is 41°10.079′N, 73°41.491′W.

p. 49. Finch's Rock House: This rock house is described as a large cave near Windmill Farm that was excavated in 1901 by archaeologists from the Museum of Natural History. The cave was named for Hiram Finch, a former property owner who died in 1897. In the early 1900s the land, as well as other properties, were acquired by Elijah Watt Sells, who named his estate North Castle Farm. Since then, North Castle Farm has been turned into a community of homes called Windmill Farm. The general GPS coordinates for North Castle Farm are 41°08.859′N, 73°40.858′W.

In volume 38, issue 1 (January–March 1962) of *The Westchester Historian of the Westchester County Historical Society*, a photograph of Finch's Rock House taken by Mary Andrews can be seen on the cover. The rock house is described as being near Armonk.

p. 49. Fishing Rock: This rock is said to be located on Fox Island at the mouth of the Byram River by Port Chester. We have been unable to locate Fox Island, but the mouth of Byram River is at a GPS location of 40°59.305′N, 73°39.412′W.

p. 59. Great Stone at the Wading Place: The rock is reputed to be located in Port Chester between New York and Connecticut, originally serving as the dividing line between Rye, New York, and Greenwich, Connecticut, that was set in 1673.

Additional information is provided by a Hope Farm website: "The survey of 1684 had begun at the mouth of the Byram River, at a point 30 miles from New York, had followed that stream as far as the head of tidewater, or about a mile and a half from the Sound, to a certain 'wading-place,' where the common road crossed the stream at a rock known and described as 'The Great Stone at the Wading-Place.'"

p. 122. Rock-shelter: This unnamed rock-shelter is a "16 meter [52.5 feet] overhanging cliff used as a permanent place of shelter for Indians." It is located in North Castle, east of Middle Patent Road near Mianus River Road.

p. 135. Spindle Rock: This unique rock, shaped like a spindle, is located in New Rochelle between Davenport Neck and Douglas Island (probably David Island), supposedly visible from Davenport Park (40°53.773'N, 73°46.275'W).

p. 143. Tobys Cave: The cave acquired its name from a Tory slave named Toby who hid there during the American Revolutionary War. The rock formation is said to be located in Rye at Parson's Woods southeast of the intersection of Theodore Fremd Avenue and North Street.

4. Emmet's Cave: William J. Myles and Daniel Chazin, *Harriman Trails: A Guide and History*, 4th ed. (New York: New York-New Jersey Trail Conference, 2018), 318.

Emmett's Cave is said to be on the east side of Brundige Mountain not far from Baker Camp. The GPS coordinates for Baker Camp at Lake Sebago are 41°12.101'N, 74°08.141'W.

The cave is named for Auntie Emmet, a witch who lived in the cave and who, according to folklore, could transform herself into a toad.

Around 1900, the body of a man named Conklin, who had been robbed and murdered, was discovered hidden in the cave. This tale may have some truth to it.

5. Balancing Rock: Stella Green and H. Neil Zimmerman, *50 Hikes in the Lower Hudson Valley* (Woodstock, VT: Backcountry Guides, 2002), 146.

The authors mention two different balancing rocks along the blue-blazed Timp-Torne and red-on-white Ramapo-Dunderberg Trails as you head west. Trailhead parking is located at a GPS location of 41°16.888′N, 73°57.752′W. A couple of websites recounting this hike make no mention of either balancing rock, which leads us to believe that either they are not all that memorable or someone has toppled them.

6. Large Boulders: Peter Senterman, "Discover and Explore the New Andes Rail Trail," *Kaatskill Life* 29, no. 1 (Spring 2014).

On page 57 is a photograph of the author standing next to "some interesting boulders split from the ledge." The New Andes Rail Trailhead is at a GPS location of 42°05.378′N, 74°49.191′W.

7. Randall Comfort, comp., *History of Bronx Borough: City of New York* (New York: North Side News Press, 1906).

p. 2. "Overlooking the new Jerome Park Reservoir, just in front of the engineer's office, stands another immense rock." The GPS reading for the general area by the reservoir is 40°52.903′N, 73°53.631′W. We suspect that the rock may no longer exist. The area, quite frankly, looks pretty developed.

p. 2. Comfort mentions a "Great Rock near the southerly limit of Claremont Park." Unfortunately, the Claremont Park website makes no mention of a large rock. For what it's worth, the GPS coordinates for the southern part of Claremont Park are 40°50.254′N, 73°54.499′W.

p. 2. "A large boulder stands near the corner of Southern Boulevard and Homes Street." The GPS reading for this junction is 40°49.712′N, 73°53.512′W. Even in Comfort's time, over 120 years ago, there was talk about the rock being destroyed to make room for the city continuing to expand. We don't see the boulder on Google Earth at the junction listed, so perhaps it was destroyed as the city expanded in size. If it still exists, it is well hidden.

p. 65. *Devil's Stepping Stones* – "Just this side of Eastchester, among the rocky fields, stands a huge boulder deeply marked with the impression of the right human foot." This potholed rock formation will be extremely difficult to find without specific directions, of which we have none to give.

8. Anita Inman Comstock, *Wondrous Westchester: Its History, Landmarks, and Special Events* (Mount Vernon, NY: Effective Learning, 1984).

Two interesting rocks are listed on page 43:

Jimmy-Under-the-Rock: A fellow named Jimmy came across a long rock projecting from the mountainside in North Castle and used it as the roof for his house. The house no longer exists, but remnants of the rock still do.

In her article "Jimmy the Rock," that appeared in the *Westchester Historian: The Quarterly of the Westchester County Historical Society* (vol. 32, no. 1, January 1956), Allison Albee suggests that Jimmy's last name may have been Johnson, and that while he was "passing through a ravine he observed an overhanging rock to one side which projected out about 20 feet from the bank and beneath which there was a space some 15 feet high." It soon became his home.

Albee states that the rock is located 0.5 mile south of the Kensico Dam and 0.25 mile east of Route 22 in Valhalla. This would place it roughly at a GPS location of 41°03.846′N, 73°45.590′W, where, coincidentally, there does appear to be a rocky area.

In *It Happened in Old White Plains*, Renoda Hoffman writes on page 66, "The large flat rock, jutting from the hillside, became his roof. He gathered the plentiful stones that lay handily about to build thick walls on three sides to shelter his shallow cave."

Kettle Hole Rock: Anita Inman Comstock contends that this hole, located in Valhalla, was "either formed by water dripping and making a small entrance at top, or by Indians chopping it."

9. Golden Bridge Balanced Rock: Maureen Koehl, *Lewisboro*, Images of America Series (Charleston, SC: Arcadia Publishing, 1997), 21.

The Goldens Bridge Balanced Rock is located in a wooded area off Route 138 in Goldens Bridge. It is found on a hillside a short distance from a stream. From the photograph displayed in the book, the rock looks to be 3–4 feet high, and longer than its height. It seems to be resting on two mounds of rock, much like a slab between two sawhorses. The brief directions given are really insufficient to even know where to begin looking for this rock.

10. Upright Rock: Maureen Koehl, *Remembering Lewisboro*, New York (Charleston, SC: The History Press, 2008), 34–44.

On page 44 is a photograph of an "upright rock nine feet high, that tradition bequeaths to us as the point reached by the Indians in claiming their lands as far as they could walk in a day starting from Stamford."

The rock is reputedly at the south end of Elmwood Road, possibly near where Elmwood Road and Smith Ridge Road intersect. The GPS reading

72. Stissing Mountain Meteor. Courtesy of the Little Nine Partners Historical Society.

for the junction of these two roads is 41°13.3898′N, 73°31.055′W. We did notice what looked like a boulder behind a private home (41°13.445′N, 73°31.116′W) just northwest of the junction, but who knows.

11. Stissing Mountain Meteor: Joyce C. Ghee and Joan Spence, *Harlem Valley Pathways: Through Pawling, Dover, Amenia, North East, and Pine Plains*, Images of America Series (Charleston, SC: Arcadia Publishing, 1998), 94.

On page 94 is a photograph of the Stissing Mountain Meteor, which looks like a 15–20-foot-high boulder. We have no specific information on this rock, but it would prove helpful if the boulder was contained within the 590-acre Stissing Mountain Multiple Use Area (41°56.4048′N, 73°41.686′W), for then, at least, access would not be an issue.

12. Ira K. Morris, *Morris's Memorial History of Staten Island, New York*, vol. 1 (New York: Memorial Publishing, 1898), 372–373.

Morris's book describes a number of rocks that we have been unable to locate:

Seal Rocks is "the name of several drift boulders at Princess Bay under Light House Hill, on which seals are occasionally seen in winter." The GPS reading for Lighthouse Hill is 40°30.459′N, 74°12.808′W.

Nigger-Head Rock is "a large boulder at the foot of the bluff at Light House Hill, Prince's Bay, and known as a landmark among fishermen." Presumably, the rock has been given a newer, less offensive name.

Strawberry Rock "received its name from the circumstance that strawberries once grew about it before the shore had washed away." The rock was located offshore near the foot of Central Avenue, Tottenville. Central Avenue appears to end before the shore, which makes us wonder if the area where the rock was located has been filled in.

Split Rock is "A large split rock seen at very low tide off the shore at the foot of Hannah Street, Thompkinsville." The shoreline is built up today. There may be no rock to view. The GPS reading at the end of Hannah Street is 40°38.187N, 74°04.384′W.

13. Margaret Lundrigan and Tova Navarra, *Staten Island*, Images of America Series (Charleston, SC: Arcadia Publishing, 1997).

Large Boulder: On page 124 is a photograph taken by Alice Austen of a girl holding a basket next to a large boulder. No specific information is provided except that we are left to conclude that the boulder must be on Staten Island.

14. Herbert B. Nichols, *Historic New Rochelle* (New Rochelle, NY: Board of Education, 1938).

p. 109. Mention is made of "an "enormous rock" between North Avenue and Carlton Terrace. The apex of the two roads is at 40°56.121'N, 73°47.536'W. That might make a good starting point if you wish to go out and look for the rock.

p. 112. Nichols states that "fine and interesting potholes can be seen in the rocks near the sound in Larchmont Shore Park." It seems entirely within the realm of possibility that these potholes can be found by anyone willing to commit some time and energy since the area covered by the park is a relatively small one.

p. 112. A glacial boulder is reputedly located "on the Shore Road at the New Rochelle-Pelham boundary line."

15. Chester A. Smith, *Peekskill, A Friendly Town, Its Historic Sites and Shrines: A Pictorial History of the City from 1654 to 1952* (Peekskill, NY: Friendly Town Association, 1952), 380.

Indian Mill (pothole): Two photographs are shown of an "Indian Mill" pothole near Indian Lake. The GPS coordinates for Indian Lake are 41°22.438'N, 73°53.218'W. That's as far as we can take you.

16. Devah and Gil Spear, eds. and comps., *The Book of Great Neck* (Great Neck, NY: Archival Reprint Company 1928), 68.

Saddle Rock: The writers indicate that "Saddle Rock lies between Great Neck Estates and Kings Post, overlooking Little Neck Bay." According to Wikipedia, "The Village of Saddle Rock is so named for an offshore boulder that gives the appearance of a saddle, first noted on a map in 1658."

Perhaps there are two different rocks, both named Saddle Rock, one offshore and one overlooking the bay.

17. Larry Penny, "Nature Notes: Glacial Erratics," *East Hampton Star*, January 18, 2012

Joshua's Rock: We presume that Joshua's Rock is located somewhere in East Hampton. There is a Joshua's Hole and a Joshua's Path in East Hampton, but that's about as much as we know.

18. Lionhead Rock: Larry Penny, "Nature Notes: Glacial Erratics," *East Hampton Star*, January 18, 2012. The rock, possibly named for Lion Gardiner, is located in Gardiners Bay (41°06.435′N, 72°12.512′W).

19. Wallace Bruce, *The Hudson: Three Centuries of History, Romance, and Invention*, centennial ed. (New York: Walking News, 1982), 58.

A-mac-lea-sin Rock: Wallace Bruce writes that "at the mouth of the Nepperhan [now called the Sawmill River] west of the creek is a large rock, called A-mac-lea-sin, the great stone to which the Indians paid reverence as an evidence of the permanency and immutability of their deity."

20. Fort Shinnecock Boulder: According to Jeremy Dennis, on his website jeremynative.com/onthissite/listing/fort-shinnecock, "This sacred glacial erratic marks the location of what may have been both the Shinnecock Fort and June Meeting location in the Shinnecock Hills. There have been many references to a contact-period Shinnecock fort, but the specific location has likely been disrupted by development."

It seems pretty likely that the boulder no longer exists. The GPS coordinates for Old Fort Pond are 40°52.790′N, 72°26.400′W.

21. John McNamara, *History in Asphalt: The Origin of Bronx Street and Place Names* (Bronx, NY: Bronx County Historical Society, 1984), 295.

p. 250. Bass Rock, aka Bess Rock, is mentioned in an 1835 deed as being the boundary marker for William Bayard's farm, which is now part of Pelham Bay Park. We guess that it may be near Stadium Avenue, but that's all it is: just a guess.

p. 295. Indian Rock is described as a rocky hummock that was "still visible at the end of Blackrock Avenue near Brucker Boulevard and Soundview Avenue up to 1965." No mention is made about what happened to Indian Rock after that.

p. 476. Sheepspen Rocks was a cluster of rocks on Tallapossa Point that formed an enclosure which to some resembled the shape of a sheep's pen. Unfortunately, these historic rocks were buried under a landfill in 1963.

The name "Tallapoosa" comes from the Tallapoosa Club, many of whose members had fought in Tallapoosa, Georgia, during the American Civil War. Tallapossa Point early on was an island in Eastchester Bay until it was joined to the mainland. It now serves as a bird habitat.

p. 440. Pigeon Rock: At one time, Pigeon Rock was a prominent rock along the Bronx River that was well known to West Farms and Van Nest youths. The rock disappeared around 1935 when the Bronx River was slightly diverted, resulting in the boulder being buried under a landfill. If you look downstream from the East Tremont Avenue Bridge spanning the Bronx River, you should be gazing at the general area where the rock once was.

22. Stella Green and H. Neil Zimmerman, *50 Hikes in the Lower Hudson Valley* (Woodstock, VT: Backcountry Guides, 2002), 128.

Large Erratic: The authors mention a "large glacial erratic to the left" while hiking the Rockhouse Mountain Loop Trail. The Rockhouse Mountain Loop trailhead (41°15.189′N, 74°03.981′W) begins across the road from a small parking area off Tiorati Brook Road (to your right).

23. Patricia Edwards Clyne, *Hudson Valley Trails and Tales* (Woodstock, NY: Overlook Press, 1990), 162.

Clyne briefly mentions two rock-shelters in Westchester County: one on the western slope of Bull's Hill (at Haines Road in Bedford Hills), and the other that was formed by fallen boulders on Hillcrest Drive, north of Briarcliff Manor. We have a few thoughts about these two rock-shelters:

Bull's Hill Rock-Shelter: GPS coordinates of 41°14.313′N, 73°42.419′ take you to the west side of Bull's Hill by Haines Road, where possibly the rock-shelter might be located.

Talus Cave: The general GPS reading for Hillcrest Drive north of Briarcliff Manor is 41°09.838′N, 73°49.188′W. There are woods to the right and possibly jumbles of rocks.

24. Harry T. Cook, *The Borough of the Bronx, 1639–1913* (New York: printed by the author, 1913), 66.

Boar's Den: Although this shelter cave was known about in the early 1900s, we can find no reference to it in the modern literature. We presume it's located somewhere in Bronx Park, perhaps near the Bronx River.

Indian Cave: According to Cook on page 106, "Perhaps the most interesting is the 'Indian' which is located a short distance east of the Hunt burying ground, and about three hundred yards north of the bridge crossing the creek." On page 107 is a photograph of the cave. The GPS reading for Hunt's Point is 40°48.705′N, 73°52.892′W.

There may be no way to track down this Indian Cave's location. In 1899, one of the Hunt family members wrote an impassioned letter to the local newspaper imploring that a projected street railway through the Hunt Burying ground be abandoned. We suspect that the plea went unanswered.

Hunt's Point today is so heavily developed that little green space remains, save for a few parks and a tiny Indian gravesite in the Joseph Rodman Drake Park (40°48.618'N, 73°52.949'W).

25. Balancing Turtle Rock

On Graham Hancock's website, grahamhancock.com/kreisberg97, a photograph of a rock called the Balancing Turtle Rock on Marlboro Mountain is shown, located in the Turtle Rock Ridge complex.

26. Robber Rocks

In *Town of Pawling: 200 Years, 1788–1988*, published by the Town of Pawling 200th Anniversary Committee, the authors write that "near a portion of the turnpike close to the top of West Mountain was the area known as 'Robber Rocks.' Here, in caves and rock piles, robbers awaited drovers returning from market ready to relieve them of their money." This rock formation is reputedly along today's Route 55.

27. Glacial Rock: Stanley H. Benham Sr., *Rural Life in the Hudson River Valley, 1880–1920* (Poughkeepsie, NY: Hudson House, 2005), 2.

A photograph taken by Sidney S. Benham of a 10–12-foot-high glacial rock in Dover Plains can be seen. No directions or hints on how to get to the boulder are given.

28. Miscellaneous Rocks: E. M. Ruttenber and L. H. Clark, *History of Orange County, New York*, vol. 1 (Interlaken, NY: Heart Lake, 1980), 34.

The authors briefly mention Kidd's Pocket-book Rock, Lover's Rocking Stone, and Poised Rock, but give no specific information about what the rocks are and where they can be found. At best, we can assume that they are in Orange County.

29. Man of War Rock: Seeley E. Ward, "Recollections of the Sloatsburg Area," *Orange County Historical Society Publication*, no. 2 (1972–1973).

On page 19, Ward writes, "Going on up above Tuxedo just a little bit, you all know where Tuxedo Station is, where the road starts to swing around a rock wall, or rocky faced hill: there was a great big rock and it was called 'Man-of-War Rock.'" A dark picture of the rock (or bluff) is shown on page 20.

30. Glacial rock: Patterson Historical Society, *Vignettes of Patterson's Past* (Patterson, NY: Patterson Historical Society, 2007).

On page 12 is a photograph with only a fraction of the rock shown. The caption reads, "Nancy Clark supports a boulder the glacier left behind on the bedrock near Route 164." It is virtually impossible to estimate the size of the boulder from the photograph. A general GPS location for Route 164 in Patterson is 41°29.047'N, 73°37.261'W.

31. Wallace Bruce, *The Hudson: Three Centuries of History, Romance, and Invention*, centennial ed. (New York: Walking News, 1982), 54.

"Among the Hudson rocks are several "'Lady's Chairs', 'Lover's Leaps', 'Devil's Toothpicks', 'Devil's Pulpit.'" No directions are given, however, on how to find these rock formations.

32. Siwanoy Indian Image Stone: Charles Dunlap, "The Image Stone of the Siwanoys," *Westchester Historian: The Quarterly of the Westchester County Historical Society* 4, no. 2 (April 1928).

Dunlap writes that the rock resembles a great bird. The stone is 2 feet high and 5 feet by 35 feet in breadth at its base, which means that it is short and wide.

33. Indian Image Stone: Ernest Freeland Griffin, ed., *Westchester County and Its People*, vol. 1 (New York: Lewis Historical Publishing, 1946).

A photo of the Indian Image Stone is shown on page 95. It is located on the grounds of the Huguenot and New Rochelle Historical Association, at the corner of North Avenue and Paine Avenues in New Rochelle. The GPS reading for the junction of these two avenues is 40°56.079'N, 73°47.504'W.

34. Old Poker Hole: Howard DeVoe, "Pleasantville's Lost Cavern," *Westchester Historian of the Westchester Quarterly* 33, no. 3 (July–September 1957), 72–74.

Mention is made of the Old Poker Hole (a cave), but no specifics are given.

The cave is also mentioned in an article by Randall Comfort entitled "The Old Kettle Hole," which he wrote for the *Westchester County Magazine* in June of 1914. The entrance is described as being somewhere along the western slope of a small, wooded ravine, south of Pleasantville's Banks Cemetery. The GPS coordinates for Pleasantville's Banks Cemetery are 41°08.263'N, 73°46.605'W. The cemetery is 0.2 mile from the junction of Route 141/Broadway and Bedford Road, going south on Route 141/ Broadway. In the late 1950s, a group of explorers looked for the cave but were unable to find it. Whether anyone has had success since then is unknown to us.

However, if you go to look for the cave, you may find something interesting along the way.

DeVoe also writes about "two interesting six foot deep potholes [that demonstrate] the existence of vigorous ground water activity at some ancient epoch."

35. Ossining Rock Shelter: Leslie V. Case, "The Ossining Rock Shelter," *Quarterly Bulletin of the Westchester County Historical Society* 75, no. 4 (October 1929): 81–85.

According to Leslie V. Case, "The shelter is formed by a great monolith torn by glacial action from the cliff of Fordham gneiss above it." It is 11 feet by 16 feet, and 6 feet at its highest point, lying 100 feet above the 9-mile-long Pocantico River. It is the smallest of three rock-shelters in close proximity. The larger two are said to have been frequented by the mysterious Leatherman.

The directions given, which are fairly imprecise, state that this rock-shelter is "½ mile south of Echo Lake; around midway between Saw Mill River Road and the Bronx Parkway Extension; near intersection of Townships of Ossining, Mt. Pleasant, and New Castle." For what it's worth, the GPS location for Echo Lake is 41°10.799'N, 73°48.609'W if you wish to use that as a starting point.

36. Leatherman Cave: Allison Albee, "The Leather Man's Cave and Washburn Mill," *Quarterly Bulletin of the Westchester County Historical Society* 30, no. 2 (April 1954): 70.

Albee writes, "Now partially destroyed, it was originally a favorite habitat of the red man. Ideally situated facing a small pond, its chamber originally measured 15 x 11 x 6 feet high."

This puts the rock-shelter in the area of Kinderogen Lake (41°09.461′N, 73°48.520′W), which can be reached from the Taconic State Parkway by getting off at Exit 6, following Pleasantville Road for 0.1 mile, and then heading northeast on Washburn Road for 1.0 mile.

37. Spook O Hole: Amy Ver Nooy, "The Ghost at Fiddler's Bridge and Other Spooks," *Year Book Dutchess County Historical Society* 42 (1957): 4, 42.

The cave, located south of Poughkeepsie, was first indicated on the Beers, Ellis, and Soule Atlas in 1867. In 1870, Spook O Hole was described in the *Poughkeepsie Telegraph* as "a remarkable cavity in a rocky hill on the property of J. and I. Frost." That was the last favorable press the cave received. In 1879, a report was made to the Poughkeepsie Society of Natural Science by J. H. Booth, A. P. Jeanarett, and Henry Booth: "This cave [Spook O Hole], if so insignificant a fissure may be dignified by such a title, is situated in Poughkeepsie Township, and lies near Barnegat, about 60 rods [990 feet] east from the Hudson River. It is in the Barnegat limestone, contains no stalactites, is very damp and will repay no one the trouble of a visit."

All of this is really academic today, for the cave in all likelihood no longer exists, having been destroyed by the enormous, nearly 2.0-mile-long, >0.5-mile-wide Clinton Point Quarry as it expanded.

38. Serpentine Cave, Indian Cave, and Prop Rock: Raymond H. Torrey, Frank Place Jr., and Robert L. Dickinson, *New York Walk Book*, 3rd ed. (New York: American Geographical Society, 1951), 15.

All that is said is that "the trail crosses Todt Hill Road and then follows bridle paths to a 'serpentine cave.'"

Dorothy Valentine Smith, a Staten Island historian, talks about a cave that she knew as Indian Cave, which was located on Todt Hill. The GPS location for Todt Hill is 40°36.094′N, 74°06.254′W.

Prop Rock: In the Black Rock Forest on a path off the Eagle Cliff Trail is a summit called Prop Rock. "There a boulder about six feet in diameter holds up, at one end, a great slab twenty feet long." As far as we can ascertain, this rock is west of Far Spy Rock (Spy Rock), north of Jim's Pond, and south of Tamarack Pond. GPS coordinates of 41°23.392′N, 74°01.540′W give you a general idea of the area where Prop Rock is.

39. Skedaddle Rock: Marjorie Smeltzer-Stevenot, *Footprints in the Ramapos* (Ashland, OH: Bookmasters, 1993), 97.

"A few who refused to be drawn into the [Civil] war hid out in 'Skedaddle Rock,' near Green Swamp in Pine Meadows, fed and protected from authorities by their families." We have no idea where this rock is, and if it really is a rock-shelter, but the Pine Meadow area is located between Johnsontown and Ladentown.

40. Peddlers Rock: Malcolm J. Mills, *East Fishkill*, Images of America Series (Charleston, SC: Arcadia Publishing, 2006), 46.

"This massive boulder, which projects into the roadway at Shenandoah and Hortontown Roads, is known locally as Peddlers Rock." We've looked for this rock at a GPS location of 41°30.995′N, 73°47.416′W, which places it right at the junction of these two roads. Unfortunately, no boulder is evident using Google Earth unless it remains hidden in the trees and brush.

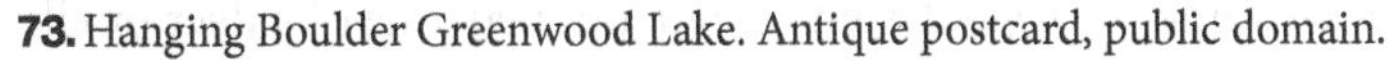

73. Hanging Boulder Greenwood Lake. Antique postcard, public domain.

41. Spring House Rock Shelter and Indian Petroglyph: Edward J. Lenik, *Indians in the Ramapos: Survival, Persistence and Presence* (North Jersey, NJ: North Jersey Highlands Historical Society, 1999), 34–37.

"The Spring House Rockshelter is an overhanging rock outcrop that protrudes from a steeply sloping hillside near the bottom of a small ravine. The shelter measures twenty feet (6 meters) in length, thirteen feet (4 meters) in depth and six and one-half feet (2 meters) in maximum height." The shelter faces west/northwest, north of Eagle Valley Road in the village of Sloatsburg. We suspect it is an archaeologically sensitive site, having been occupied by Native Americans from ca. 5,000 B.C. to ca. 1680. Artifacts from the rock-shelter are on permanent exhibit at the Sloatsburg Public Library (41°09.361'N, 74°011.615'W) on 1 Liberty Rock Road in Sloatsburg.

Lenik also writes that "in 1990, an Indian petroglyph (rock carving) was discovered on an outcrop of bedrock located on the Hemlock Hill Trail in Harriman State Park, New York. Two incised designs were carved on the top surface of the bedrock within the old hiking trail situated south of Tom Jones Mountain." Two designs are mentioned: one of a deer or elk, and one of a bear. Lenik estimates that they were carved between 1,000 A.D. and 16,000 A.D.

42. Hanging Boulder

In an early, ca. 1900 postcard, Hanging Boulder is shown perched next to a one-lane road in the Mount Peter area of Greenwood Lake. Given the passage of over a century of time, it seems unlikely that this precariously balanced rock still exists, but who knows for sure until you go to take a look.

43. Balanced Rock: William Thompson Howell, *The Hudson Highlands*, vol. 1 (A memorial set of books published in 1933, limited to 200 copies).

On page 54, the author writes that the Balanced Rock on Bear Mountain "was fixed, but in a way that to say the least was unusual. It may have weighed, perhaps, fifty tons (estimate of G. W. P.), and it rested on three points, each of which was no larger than two good sized fists. One of these points was the bed rock, and the other two were cobblestones. They raised the boulder clear of the bedrock by about six inches." Without specific directions, this would be a hard boulder to find on an area encompassed by a mountain.

44. William Thompson Howell, *The Hudson Highlands*, vol. 2 (A memorial set of books published in 1933, limited to 200 copies).

Several rocks, without directions, are mentioned by the author:

p. 9. "Bear's den, below Summit Rock on the West Point Road."

p. 17. "On top of the Profile, over fourteen hundred feet up, is a small boulder which formerly rocked at the touch. It has lately fallen over and is now fixed."

p. 17. "One [boulder] weighing a ton or more, on top of Echo Rock Ridge, at nearly as high an elevation [1,400 feet], is so balanced as to rock with a swing of several inches."

p. 17. "Behind Mine Hill and over a thousand feet up are some very large boulders for that height and locality."

45. St. Anthony's Nose: William J. Blake, *History of Putnam County, New York* (New York: Baker and Scribner, 1849), 165–167.

At one time there was a rock face on Breakneck Mountain named St. Anthony's Face, which is where St. Anthony's Nose comes from: "'St. Anthony's Face,' so celebrated in the history of the Hudson scenery, once peered out and over the rocky battlements below, gazing, as it were, at the eternal ebb and flood of the mighty current [Hudson River]. . . . In the summer of 1846, Capt. Deering Ayers, who was engaged in the services of Harlem High Bridge Company, by one fell blast, detached an immense block of granite weighing nearly two thousand tons"—and so, another geological formation of note in the mid-nineteenth century ceased to exist.

It is said that the rock formation was named for Captain Antony Hogans.

46. Raymond H. Torrey, Frank Place Jr., and Robert L. Dickinson, *New York Walk Book*, 3rd ed. (New York: American Geographical Society, 1951).

p. 99. "On the east side of the spur of Brundige Mountain, about a mile north of Burnt Saw Mill Bridge, is a perpendicular crack of 3 or 4 feet in width and extending upward for 10 or 25 feet." A pen-and-ink drawing of the opening can be seen on page 120.

p. 121. "Washington Rock is a projecting point of the trap rock ridge 520 feet above sea level . . . Here, in May and June of 1777, General Washington spent many days sweeping the country through his glass for movements of the British. This has been commemorated by a tablet erected here on a stone monument."

p. 122. "Around Chimney Rock was the encampment of the Continental troops and General Washington during the spring of 1777 . . . About Chimney Rock clings the shadowy romance of an Indian girl whose lover had been killed by a rival. She, seeking the rock where she had last seen him, hears her lover calling. She leaps from the rocks into his arms, as she thinks, and is killed on the rocks below." A pen-and-ink drawing is on page 120.

47. Stepping Stones: Robert Bolton Jr., *A History of the County of Westchester from Its First Settlement to the Present Time*, vol. 2 (New York: Alexander S. Gould, 1848), 239.

"Throckmorton's Point is likewise remarkable as the place where the tides meet in the Sound. Directly opposite are the famous *stepping stones*, a number of rocks which project in a line from the Long Island shore, and show their bare tops at low water."

48. John Dobbs Cave and Ardsley Cave: Allison Albee, *Westchester Historical Society and Golden Anniversary of the Westchester County Historical Society* 50, no. 4 (Fall 1974): 76. The article is unnamed.

Two caves (probably rock-shelters) in Yorktown are mentioned: John Dobbs Cave and Ardsley Cave. The Dobbs Cave was named for John Dobb, a Civil War deserter who lived in the cave, venturing out at night to steal food and necessities. He was spotted stealing milk early one morning by a former officer who recognized Dobbs and shot him dead.

Ardsley Cave is located somewhere in the Thirty Deer Ridge area above Jackson Avenue. This is really not a lot to go on, however.

49. Potholes: Lloyd Ultan, *The Northern Borough: A History of the Bronx* (Bronx, NY: Bronx Historical Society, 2009), 3.

Ultan states that "Other potholes can be seen near the old Bartow Station and near the bridge at Shore Road in Pelham Bay Park."

The old Barton Station (40°52.154'N, 73°48.776'W), which provided a line into City Island, stopped service in 1919, and has remained abandoned since the 1930s. Nearby is the Hutchinson River, which is where the potholes are likely to be found.

50. Fifth of July Rock, Jupiter's Boulder, and Split Rock

The first two rock formations and a Split Rock Trail show up on *West Hudson Trails Map 7: Black Rock Forest and Storm King State Park*, located in the 3,920-acre Black Rock Forest.

We had originally planned to write up a sequential route to access these rocks from a trailhead off Old Spring Road, but then learned that this entrance had been closed. The approach will now have to be from the northwest part of the Forest Preserve from Reservoir Road, off Route 9W. You will need a trail map for this one (blackrockforest.org/wp-content/uploads/2021/12/brf-trail-map-bwapril2020.pdf).

We did come up with approximate GPS coordinates for the rocks: the Fifth of July Rock, near Mineral Spring Falls, is at 41°22.840′N, 74°03.734′W (estimated); Jupiter Rock which, by name, sounds impressive, is at 41°23.043′N, 74°02.826′W (estimated); and Split Rock, overlooking Sutherland Pond, is at 41°23.699′N, 74°02.254′W (estimated) which, according to the sixth edition of the New York-New Jersey Trail Conference's *New York Walk Book*, is also known as Echo Rock.

51. Settler's Rock: This pint-sized boulder bears a plaque that reads, "Near this spot in June 1640 landed the colonists from Lynn, Mass. who founded Southampton, the first English settlement in the state of New York." The boulder (40°56.352′N, 72°25.009′W) is located at Conscience Point on land owned by the Southampton Historical Museum.

A photograph of the rock with the caption "North Sea Harbor and Rock Monument" can be seen in William Donaldson Halsey's 1935 book *Sketches from Local History*.

52. Clay Perry, *Underground Empire: Wonders and Tales of New York Caves* (New York: Stephen Daye Press, 1948).

p. 204. Money Hole Cave: Clay Perry writes that the cave is "in granite, pegmatite and gneiss about 30 by 30 feet in area from 4 to 7 feet high, and entered by a 3-foot opening, with a drop of 25 feet inside." It was inhabited by Henry Holmes around 1820, who used it as a hideout for counterfeiting money.

Patricia Edwards Clyne visited the cave in the 1970s and provided directions to it in her 1980 book *Caves for Kids* (pp. 34–45) that, unfortunately, are no longer comprehensible due to vanishing landmarks. However, in the *Poughkeepsie Journal* (April 4, 2016), John Ferro writes that the cave was rediscovered by the town of Lloyd police lieutenant James Janso after several years of hiking the woods near Indian Brook Road in Fahnestock Memorial State Park. He found two rooms, one higher than the other, with the largest room being 5 feet wide and 20 feet long. If Officer Janso was able to find the cave, then obviously you can, too, with the proper diligence.

p. 205: Continentalville Cave: The cave "was formed by a split in the pegmatite in a hillside, which formed a room, beyond a narrow opening, some fifty-five feet long and seven feet high, three feet wide." The author speculates that it may have been a Tory hideout.

It seems likely that the cave is located in the hamlet of Continentalville, today known as Continental Village, but that's as far as we can take you.

53. Dinosaur Rock: Frank L Walton, *The Cedar Knolls* (Bronxville, NY: Cedar Knolls Colony, 1960).

On page 7 is a photograph of a large rock with the caption, "Dinosaur Rock: 86 Dellwood Road." The GPS coordinates at this location, 40°56.823′N, 73°50.750′W, reveal a large mound of bedrock extending down to the sidewalk in front of a house.

54. Mucklestone Rock: Renoda Hoffman, *It Happened in Old White Plains* (White Plains, NY: printed by the author, 1981), 39.

"Not until one stands on the rock can the sheer face be seen as it falls precipitously without warning 50 feet to a glen below." *Mucklestone* is a Scottish word for "great quantities."

During the Revolutionary War, the rock was known as Tiltan's Rock after Captain Tiltan, a British officer who, being pursued by American forces, inadvertently plummeted over the top of the rock with his horse during the chase.

Hoffman mentions that this rock, or perhaps rocky mound or cliff, is east of Hall Avenue near I-287 in White Plains.

55. Giant Neighborhood Rocks in New York City: Jen Carlson, "This Gigantic Rock Jammed between Two Boulders Is a Huge Part of NYC History," *Gothamist*, last modified August 2, 2018 (gothamist.com/

arts-entertainment/this-gigantic-rock-jammed-between-two-buildings-is-a-huge-part-of-nyc-history).

The website shows a photograph of a 30-foot-high, 100-foot-long boulder on the south side of 114th Street between Broadway and Riverside Drive, wedged in between two buildings at a GPS of 40°48.413'N 73°57.930'W. The boulder and surrounding property are owned by Columbia University.

Carlson also mentions another large rock, nearly half the size of the buildings surrounding it, that is located on the west side of Bennett Avenue between 181st Street and 184th Street at a GPS of 40°51.067'N 73°56.189'W.

ACKNOWLEDGMENTS

Many thanks go to—

Dan Balogh, creator of danbalogh.com, for generously providing the majority of the photographs pertaining to hikes along the west side of the Hudson River; Richard Delaney, for proofreading an earlier version of this book; Barbara Delaney, who not only proofread an earlier version of this book, but participated in some of the hikes; Daniel Chazin, author, photographer, and hiker, for his photographs of Split Rock and Bear Rock; Alex Smoller, for his photograph of a large glacial erratic in the Betsy Sluder Preserve; Christian Prellwitz for his photograph of the large East Marion Boulder; and Steven Schimmrich, the Hudson Valley geologist, for his photograph of Turtle Rock.

Lisa Motluck for reviewing the chapter on the Walt Whitman Boulder and for her input on the elusive Bear Mountain Boulder; David Beck, park naturalist, Dutchess County, for guiding us to the North Rockledge Shelter at Bowdoin Park.

Brendan Murphy, director of stewardship, Westchester Land Trust, for helping us determine whether an accessible rock-shelter exists in the Westchester County's Rockshelter Preserve—it doesn't; Barbara Ransome, director of operations, Greater Port Jefferson Chamber of Commerce, without whose help we would not have located the Jefferson Boulder; Amanda Cymore, park ranger, and her fellow park rangers who spent considerable time on the summit of Bear Mountain looking for a boulder that we believed to be there—but, apparently, wasn't. Steve Olsen, parks director, Dutchess County; Betsy Biddle, executive director at Andrus on Hudson; and Christine Tesauro, at Port Jefferson—all for being so helpful.

We particularly want to thank the staff at the State University of New York Press for their help and wonderful suggestions in putting this book together: Ryan Morris, production and design manager; Aimee Harrison, who worked on the cover design; Julia Cosacchi, assistant manuscript editor; and Richard Carlin, senior acquisitions editor for music, education, and New York State history and culture.

ABOUT THE AUTHOR

Russell Dunn, a former New York State licensed hiking guide, lives in Albany, New York, is married to Barbara Delaney (a fellow writer), and is the author of eleven waterfall guidebooks, five paddling guidebooks, eight hiking guidebooks, *Adventures around the Great Sacandaga Lake* (his first book), *Ausable Chasm in Pictures and Story* (co-authored with John Haywood and Sean Reines), and eleven photobooks of stereographic pictures.

Dunn's hobbies, when he is not writing or exploring, are stereography, legerdemain, songwriting, and playing the guitar.

He can be reached at rdunnwaterfalls@yahoo.com.

BIBLIOGRAPHY

Adams, Arthur G. *The Hudson River Guidebook*. New York: Fordham University Press, 1996.

Adkins, Leonard M., and the Appalachian Trail Conservancy. *Along the Appalachian Trail: New Jersey, New York, and Connecticut*. Images of America Series. Charleston, SC: Arcadia Publishing, 2014.

Adler, Cy A. *Walking the Hudson, Batt to Bear: From the Battery to Bear Mountain*. Cathedral, NY: Green Eagle Press, 1997.

Albee, Allison. "Jimmy the Rock." *Westchester Historian: The Quarterly of the Westchester County Historical Society* 32, no. 1 (January 1956).

———. "The Leather Man's Cave and Washburn Mill." *Westchester Historian: The Quarterly of the of the Westchester County Historical Society* 30, no. 2 (April 1954).

———. "A Natural Bridge in Westchester." *Westchester Historian: The Quarterly of the Westchester County Historical Society* 33, no. 4 (November–December 1957).

———. *Westchester Historian: The Quarterly of the Westchester County Historical Society* 50, no. 3 (Fall 1974). Article untitled.

Andrews, Mary. *Westchester Historian: The Quarterly of the Westchester County Historical Society* 38, no. 2 (April–June, 1962).

Antos, Jason D. *Whitestone*. Images of America Series. Charleston, SC: Arcadia Publishing, 2006.

Asimov, Isaac. *The Shaping of North America*. London: Dobson Books, 1973.

Bailey, Bill. *New York State Parks: A Guide to New York State Parks*. Saginaw, MI: Glovebox Guidebooks of America, 1997.

Barrett, Kevin. *Newburgh*. Images of America Series. Charleston, SC: Arcadia Publishing, 2007.

Bayles, Richard M., ed. *History of Richmond County, (Staten Island) New York, from Its Discovery to the Present Times*. New York: L. E. Preston, 1887.

Beach, Lewis. *Cornwall*. Newburgh, NY: E. M. Ruttenber & Sons, 1873.

Beard, Robert D. *Rockhounding New York: A Guide to the State's Best Rockhounding Sites*. Guilford, CT: Falcon Guides, 2014.

Bedell, Cornelia F. *Now and Then and Long Ago in Rockland County, New York*. New City, NY: Historical Society of Rockland County, 1968.

———. *Now and Then and Long Ago in Rockland County, New York*. Rockland County, NY: privately printed, 1941.

Benham, Stanley H., Sr. *Rural Life in the Hudson River Valley, 1880–1920*. Poughkeepsie, NY: Hudson House, 2005.

Bernhard, Arthur I. "Katonah and Bedford: A Do-It-Yourself Historical Tour." *Westchester Historian: The Quarterly of the Westchester County Historical Society* 43, no. 2 (Spring 1967).

Blake, William J. *History of Putnam County, New York*. New York: Baker and Scribner, 1849.

Bolton, Reginald Pelham. *Indian Life of Long Ago in the City of New York*. New York: Bolton Books, 1924.

———. *Washington Heights, Manhattan: Its Eventful Past*. New York: Dyckman Institute, 1924.

Bolton, Robert, Jr. *A Guide to New Rochelle and Lower Westchester*. Harrison, NY: Harbor Hill Books, 1976 facsimile of the 1842 book.

———. *A History of the County of Westchester, from Its First Settlement to the Present Time*. 2 vols. New York: Alexander S. Gould, 1848.

Brown, Henry Collins, ed. *Valentine's Manual of Old New York*. New York: Valentine's Manual, 1927.

Bruce, Wallace. *The Hudson: Three Centuries of History, Romance, and Invention*. Centennial ed. New York: Walking News, 1982.

Bryson, J. "Rock Hill, Long Island, N.Y.," *American Geologist* 16 (1895).

Case, Daniel. *AMC's Best Day Hikes near New York City*. Boston: Appalachian Mountain Club, 2010.

Case, Leslie V. "The Ossining Rock Shelter." *Quarterly Bulletin: Westchester County Historical Society* 75, no. 4 (October 1929).

Chong, Herb, ed. *The Long Path Guide*. 5th ed. Mahwah, NJ: New York-New Jersey Trail Conference, 2002.

City of New York Parks and Recreation. *Pelham Bay Park History*. New York: Administrator's Office, City of New York Parks and Recreation, 1986.

Clayton, W. Woodford, and William Nelson, comps. *History of Bergen and Passaic Counties, New Jersey: With Biographical Sketches of Many of Its Pioneers and Prominent Men*. Philadelphia: Everts and Peck, 1881.

Clyne, Patricia Edwards. *Caves for Kids in Historic New York*. Monroe, NY: Library Research Associates, 1980.

———. *Hudson Valley Faces and Places*. Woodstock, NY: Overlook Press, 2005.

———. *Hudson Valley Trails and Tales*. Woodstock, NY: Overlook Press, 1990.
Comfort, Randall, comp. *History of Bronx Borough: City of New York*. New York: North Side News Press, 1906.
Comstock, Anita Inman. *Wondrous Westchester: Its History, Landmarks, and Special Events*. Mount Vernon, NY: Effective Learning, 1984.
Cook, Harry T. *The Borough of the Bronx, 1639–1913: Its Marvelous Development and Historical Surroundings*. New York: printed by the author, 1913.
Cornell, Greta. "Frank's Rock." *Westchester Historian: The Quarterly of the Westchester County Historical Society* 41, no. 1 (1965).
Day, Leslie. *Field Guide to the Natural World of New York City*. Baltimore, MD: Johns Hopkins University Press, 2007.
DeVoe, Howard. "Pleasantville's Lost Cavern." *Westchester Historian: The Quarterly of the Westchester County Historical Society* 33, no. 3 (July–September 1957).
Diamant, Lincoln. *Teatown Lake Reservation*. Images of America Series. Charleston, SC: Arcadia Publishing, 2003.
Dolkart, Andrew S. *Guide to New York City Landmarks*. New York: John Wiley & Sons, 1998.
Dolkart, Andrew S., and Gretchen S. Sorin. *Touring Historic Harlem: Four Walks in Northern Manhattan*. New York: New York Landmark Conservancy, 1997.
Duncombe, Frances R., and the Historical Committee, Katonah Village Improvement Society, *Katonah: The History of a New York Village and Its People*. Katonah, NY: The Society, 1961.
Dunlap, Charles. "The Image Stone of the Siwanoys." *Westchester Historian: The Quarterly of the Westchester County Historical Society* 4, no. 2 (April 1925).
Dunn, Russell. *Hudson Valley Waterfall Guide*. Hensonville, NY: Black Dome Press, 2005.
Dunn, Russell, and Barbara Delaney. *Paths to the Past: History Hikes through the Hudson River Valley, Catskills, Berkshires, Taconics, Saratoga and Capital Region*. Catskill, NY: Back Dome Press, 2021.
Eichner, Frances, and Helen Ferris Tibbets, eds. *When Our Town Was Young: Stories of North Salem's Yesterday*. North Salem, NY: Town of North Salem, 1945.
"Elastic Cave." *Westchester Historian: The Quarterly of the Westchester County Historical Society* 31, no. 2 (April 1955).
Farkas, Diana. "The Zoological Park, Bronx, NY." *Bronx County Historical Society Journal* 11, no. 1 (Spring 1974).
Feirstein, Sanna. *Naming New York Manhattan: Places and How They Got Their Names*. New York: New York University Press, 2001.
Flad, Harvey K., and Clyde Griffin. *Main Street to Mainframes: Landscape and Social Change in Poughkeepsie*. Albany, NY: State University of New York Press, 2009.

Flynn, Kevin. "A Hike into the Mystic, or Just a Walk in the Woods?" *New York Times*, October 15, 2010. https://nytimes.com/2010/10/15/nyregion/15hawk.html.

Freund, James. *Central Park: A Photographic Excursion*. New York: Fordham University Press, 2001.

Funk, R. E., and D. W. Steadman. *Dutchess Quarry Caves, Orange County, New York*. N.p.: Persimmon Press Monographs in Archaeology, 1994.

Gekle, William F. *The Lower Reaches of the Hudson River*. Poughkeepsie, NY: Wyvern House, 1982.

Gethard, Chris. *Weird New York: Your Travel Guide to New York's Local Legends and Best Kept Secrets*. New York: Sterling, 2005.

Ghee, Joyce C., and Joan Spence. *Harlem Valley Pathways: Through Pawling, Dover, Amenia, North East, and Pine Plains*. Images of America Series. Charleston, SC: Arcadia Publishing, 1998.

——. *Poughkeepsie, 1898–1998: A Century of Change*. Images of America Series. Charleston, SC: Arcadia Publishing, 1999.

——. *Poughkeepsie: Halfway up the Hudson*. Images of America Series. Charleston, SC: Arcadia Publishing, 1997.

Gottlock, Barbara H., and Wesley Gottlock. *New York's Palisades Interstate Park*. Images of America Series. Charleston, SC: Arcadia Publishing, 2007.

Graff, M. M. *Central Park, Prospect Park: A New Perspective*. New York: Greensward Foundation, 1985.

Gray, Christopher. "Central Park Indian Cave," *Northeastern Caver* 42, no. 3 (September 2011): 89.

"Great Stone Face Rock Shelter." *Quarterly Bulletin of the Westchester County Historical Society* 7, no. 1 (January 1931).

Green, Stella, and H. Neil Zimmerman. *50 Hikes in the Lower Hudson Valley*. Woodstock, VT: Backcountry Guides, 2002.

Griffin, Ernest Freeland, ed. *Westchester County and Its People*. Vol. 1. New York: Lewis Historical Publishing, 1946.

Haagensen, Alice Munro. *Palisades and Snedens Landing*. Tarrytown, NY: Pilgrimage Publishing, 1986.

Halsey, William Donaldson. *Sketches from Local History*. Bridgehampton, NY: n.p., 1935.

Hansen, Harry. *North of Manhattan: Persons and Places of Old Westchester*. New York: Hastings House, 1950.

Harmon, Chris, Matt Levy, and Gabrielle Antoniadis. *Eastern New York Chapter Preserve Guide: Lower Hudson Region*. Mt. Kisco, NY: Nature Conservancy, 2000.

Hasbrouck, Frank. *The History of Dutchess County, New York*. Poughkeepsie, NY: S. A. Mattheu, 1909.

"The Hermitess of Salem." *Westchester Historian: The Quarterly of the Westchester County Historical Society.* 19, no. 1 & 2. (January–April 1943).
Herr, Beth, and Maureen Koehl. *Ward Pound Ridge Reservation.* Images of America Series. Charleston, SC: Arcadia Press, 213.
Hine, Charles Gilbert, comp. *History and Legend of Howard Avenue and the Serpentine Road, Grymes Hill, Staten Island.* [New York?]: printed by the author, 1914.
Hinkemeyer, Arlene. *A History of the Incorporated Village of Plandome Heights.* Manhasset, NY: Incorporated Village of Plandome Heights, 1997.
Hoffman, Renoda. *It Happened in Old White Plains.* White Plains, NY: printed by the author, 1981.
Howell, William Thompson. *The Hudson Highlands.* 2 vols. N.p.: privately published, 1933.
——. *The Hudson Highlands.* 2 vols. New York: Walking News, 1982.
Hutchinson, Lucille, and Ted Hutchinson. *Storm's Bridge: A History of Elmsford, N.Y., 1700–1976.* Elmsford, NY: Bicentennial Committee, 1980.
Imbrogno, Philip J. "The Mysteries of Hawk Rock." *Greenwich Time,* September 8, 2010. https://greenwichtime.com/opinion/article/The-mysteries-of-Hawk-Rock-649692.php.
Ingersoll, Ernest. *Handy Guide to the Hudson River and Catskill Mountains.* Astoria, NY: J. C. & A. L. Fawcett, 1989; reprint of 1910 book.
Jenkins, Stephen. *The Story of the Bronx: From the Purchase Made by the Dutch from the Indians in 1639 to the Present Day.* New York: G. P. Putnam's Sons, 1912.
Jonnes, Jill. *South Bronx Rising: The Rise, Fall, and Resurrection of an American City.* New York: Fordham University Press, 2002.
Kick, Peter, Barbara McMartin, and James M. Long. *50 Hikes in the Hudson Valley: From the Catskills to the Taconics, and from the Ramapos to the Helderbergs.* 2nd ed. Woodstock, VT: Backcountry Publications, 2000.
Klein, Howard. *Three Village Guidebook: The Setaukets, Poquott, Old Field and Stony Brook.* 2nd ed. Illustrated by Patricia Windrow. East Setauket, NY: Three Village Historical Society, 1986.
Koehl, Maureen. *Lewisboro.* Images of America Series. Charleston, SC: Arcadia Publishing, 1997.
——. *Remembering Lewisboro, New York.* Charleston, SC: History Press, 2008.
Lawrence, Frances Meyer. "The Story of a Rock." *Nassau County Historical Journal* 15, no. 1 (Spring 1954).
Lederer, Richard M., Jr. *The Place-Names of Westchester County, New York.* Harrison, NY: Harbor Hill Books, 1978.
Lenik, Edward J. *Indians in the Ramapos: Survival, Persistence and Presence.* North Jersey, NJ: North Jersey Highlands Historical Society, 1999.

——. *Iron Mine Trails*. New York: New York-New Jersey Trail Conference, 1996.
Levine, Edward J. *Central Park*. Postcard History Series. Charleston, SC: Arcadia Publishing, 2006.
Levinson, Lee Ann. *East Side, West Side: A Guide to New York City Parks in All Five Boroughs*. Darien, CT: Two Bytes, 1997.
Lightfoot, Frederick S., Linda B. Martin, and Bette S. Weidman. *Suffolk County, Long Island in Early Photographs, 1867–1931*. New York: Dover Publications, 1984.
Lightfoot, Kent G., Robert J. Kalin, and James Moore. *Prehistoric Hunter-Gatherers of Shelter Island, New York: An Archaeological Study of the Mashomack Preserve*. Berkeley, CA: Archaeological Research Facilities, 1987.
Lossing, Benson. *The Hudson: From the Wilderness to the Sea*. Sommersworth, NY: New Hampshire Publishing Company, 1972; facsimile of the 1866 edition.
Lossing, Thomas Sweet. *My Heart Goes Home: A Hudson Valley Memoir*. Fleischmanns, NY: Purple Mountain Press, 1997.
Lubar, Harvey. "The History of Indian Lake," *Bronx County Historical Society* 22, no. 2 (Fall 1985).
Lundrigan, Margaret, and Tova Navarra. *Staten Island*. Images of America Series. Charleston, SC: Arcadia Publishing, 1997.
Mack, Arthur C. *The Palisades of the Hudson*. Edgewater, NJ: Palisade Press, 1909.
Martin, Edward, photographer. *Yonkers Historical Bulletin* 9, no. 1 (April 1962).
McNamara, John. *History in Asphalt: The Origin of Bronx Street and Place Names*. Harrison, New York: Harbor Hill Books, 1978.
——. *History in Asphalt: The Origin of Bronx Street and Place Names*. Rev. ed. Bronx, NY: Bronx County Historical Society, 1984.
——. *McNamara's Old Bronx*. Bronx, NY: Bronx County Historical Society, 1989.
McNew, George L. "The Paradise World of the Red Man in Westchester County." *Yonkers Historical Bulletin* 15, no. 1 (January 1965).
Merguerian, Charles, and Charles A. Baskerville. "Geology of Manhattan Island and the Bronx, New York City, New York." In *Northeastern Section of the Geological Society of America*, ed. David C. Roy, 137–149. Vol. 5 of *Centennial Field Guide*. Boulder, CO: Geological Society of America, 1987.
Mills, Malcolm J. *East Fishkill*. Images of America Series. Charleston, SC: Arcadia Publishing, 2006.
Montanarelli, Lisa. *New York City Curiosities: Quirky Characters, Roadside Oddities, and Other Offbeat Stuff*. Guilford, CT: Morris Book Publishing, 2011.
Morris, Ira K. *Morris's Memorial History of Staten Island, New York*. Vol. 1. New York: Memorial Publishing, 1898.
Mosco, Steve. "What Is Shelter Rock?" *Manhasset Press*, September 20, 2015.
Myles, William J. *Harriman Trails: A Guide and History*. New York: New York-New Jersey Trail Conference, 1994.

Myles, William J., and Daniel Chazin. *Harriman Trails: A Guide and History*. 4th ed. New York: New York-New Jersey Trail Conference, 2018.

New York-New Jersey Trail Conference. *Day Walker: 32 Hikes in the New York Metropolitan Area*. 2nd ed. Mahwah, NJ: New York-New Jersey Trail Conference, 2002.

———. *Guide to the Appalachian Trail in New York and New Jersey*. 9th ed. Harpers Ferry, WV: Appalachian Trail Conference, 1983.

———. *New York Walk Book*. 6th ed. New York: New York-New Jersey Trail Conference, 1998.

Nichols, Herbert B. *Historic New Rochelle*. New Rochelle, NY: Board of Education, 1938.

"North Salem's Great Granite Boulder." *Westchester Historian: The Quarterly of the Westchester County Historical Society* 3, no. 1 (January–March 1959).

Ochojski, Paul M., ed. *More Gleanings from Rockland History* (Orangeburg, NY: Journal of the Historical Society of Rockland County, 1971).

Orange-Ulster Board of Cooperative Education Services. *Orange County: A Journey through Time*. Orange County, NY: reprint, 1984.

Owen, James. "The Fortified Indian Village at Crown Point." *Quarterly Bulletin of the Westchester Historian* 2, no. 2 (April 1956).

Owens, William. *Pocantico Hills 1609–1959*. Sleepy Hollow, NY: Sleepy Hollow Restorations, 1960.

Palmer, Virginia. "The Stone Church." *Year Book: Dutchess County Historical Society* 33 (1948).

Patterson Historical Society. *Vignettes of Patterson's Past*. Patterson, NY: Patterson Historical Society, 2007.

Penny, Larry. "Nature Notes: Glacial Erratics." *East Hampton Star*, January 18, 2012.

Perls, Jeffrey. *Paths along the Hudson: A Guide to Walking and Biking*. New Brunswick, NJ: Rutgers University Press, 2001.

Perry, Clay. *Underground Empire: Wonders and Tales of New York Caves*. New York: Stephen Daye Press, 1948.

Peters, Mary K. "Address at the Unveiling of a Tablet to George Fox at Council Rock." *Nassau County Historical Journal* 5, no. 1 (March 1942).

Pierce, Carl Horton. *New Harlem: Past and Present*. New York: New Harlem Publishing, 1903.

The Rockland Record. Vol. 2. Rockland County, NY: Rockland County Society of the State of New York, 1931.

Rogers, Elizabeth Barlow. *Rebuilding Central Park: A Management and Restoration Plan*. Edited by Marianne Cramer, Judith L. Heintz, Bruce Kelly, Philip N. Winslow, and John Berendt. Cambridge, MA: MIT Press, 1987.

Rokach, Allen. "History Underfoot: A Short Geological History of the Bronx." *Bronx County Historical Society Journal* 11, no. 2 (Fall 1974).

Roseberry, C. R. *Niagara to Montauk: The Scenic Pleasures of New York State.* Albany, NY: State University of New York Press, 1982.

Rosenzweig, Roy, and Elizabeth Blackman. *The Park and the People: A History of Central Park.* Ithaca, NY: Cornell University Press, 1992.

Ruttenber, E. M., and L. H. Clark. *History of Orange County, New York.* Vol. 1. Interlaken, NY: Heart Lake, 1980.

Ryan, Robert F. "John Glover and the Battle of Pell's Point." *Bronx County Historical Society Journal* 2, no. 2 (July 1965).

Scharf, John Thomas. *History of Westchester County, New York.* Vol. 1. Philadelphia: L. E. Preston, 1886.

———. *History of Westchester County.* Vol. 2. New York: L. E. Preston, 1886.

Scheier, John. *New York City Zoos and Aquarium.* Images of America Series. Charleston, SC: Arcadia Publishing, 2005.

Schuberth, Christopher J. *The Geology of New York City and Environs.* Garden City, NY: Natural History Press, 1968.

Scott, Catherine A. *City Island and Orchard Beach.* Charleston, SC: Arcadia Publishing, 1999.

Scott, Catherine. "Twin Island: A Bronx Secret." *Bronx County Historical Society Journal* 35, no. 1 (Spring 1998).

Secord, Morgan. "Cat Rock Cave." *Westchester Historian: The Quarterly of the Westchester County Historical Society* 41, no. 4 (Autumn 1965).

Secord, Morgan H. "The Indian Rock Shelter of Larchmont." *Westchester Historian: The Quarterly of the Westchester County Historical Society* 38, no. 2 (April–June 1962).

Seitz, Sharon, and Stuart Miller. *The Other Islands of New York City: A History and Guide.* Woodstock, VT: Countryman Press, 2001.

Senterman, Peter. "Discover and Explore the New Andes Rail Trail." *Kaatskill Life* 29, no. 1 (Spring 2014).

Serrao, John. *The Wild Palisades of the Hudson.* Westwood, NJ: Lind Publications, 1949.

Shonnard, Frederic, and W. W. Spooner. *History of Westchester County, New York.* New York: New York History Company, 1900.

Silverman, Ann B. "Guarding County's Archaeological Past." *New York Times,* November 17, 1985.

Smeltzer-Stevenot, Marjorie. *Footprints in the Ramapos.* Ashland, OH: Bookmasters, 1993.

Smith, Carlyle S. "Manhasset Rock." Unpublished manuscript on file with the Department of Anthropology Archives at the American Museum of Natural History, June 10, 1946.

Smith, Chester A. *Peekskill, a Friendly Town, Its Historic Sites and Shrines: A Pictorial History of the City from 1654 to 1952.* Peekskill, NY: Friendly Town Association, 1952.

Smith, Julia S. "Old Setauket." *Long Island Historical Society Quarterly* 2, no. 2 (April 1940).

Smith, Philip H. *Legends of the Shawangunk and its Environs*. Fleischmanns, NY: Purple Mountain Press, 1967.

Somers Historical Society. *Somers: Its People and Places 1788–1988, a 200 Year History*. Somers, NY: Somers Historical Society, 1989.

Spear, Devah, and Gil Spear, eds and comps. *The Book of Great Neck*. Great Neck, NY: Archival Reprint Company, 1928.

"A Special Section: Central Park." *The Conservationist*. 28, no. 4 (February–March 1974).

Spikes, Judith Doolin. *Larchmont, NY: People and Places*. Larchmont, NY: Fountain Square Books, 1991.

Spinzia, Raymond E., Judith A. Spinzia, and Kathryn E. Spinzia. *Long Island: A Guide to New York's Suffolk and Nassau Counties*. New York: Hippocrene Books, 1991.

Starbuck-Ribaudo, Diane, William Corbet, and AnnMarie Fishwick. "A Study of Erratics on the Ronkonkoma Moraine in Eastport, Long Island." Unpublished manuscript, n.d. https://dspace.sunyconnect.suny.edu/bitstream/handle/1951/47814/starbuck.pdf.

Steiner, Henry. *The Place Names of Historic Sleepy Hollow and Tarrytown*. Bowie, MD: Heritage Books, 1998.

Stickney, Charles E. *A History of the Minisink Region*. Middletown, NY: Coe Finch and I. F. Guiwits, 1867.

Stiegelmaier, Kevin. *Canoeing and Kayaking New York*. Birmingham, AL: Menasha Ridge Press, 2009.

Swanson, Susan Cochran, and Elizabeth Green Fuller. *Westchester County: A Pictorial History*. Norfolk, VA: Donning, 1982.

Swayne, Lawrence C. "Revolutionary War Heroism Acknowledged." *Kaatskill Life* 22, no. 1 (Spring 2007).

Thomas, Bill, and Phyllis Thomas. *Natural New York*. New York: Holt, Rinehart and Winston, 1983.

Three Village Historical Society. *The Setaukets, Old Field, and Poquott*. Images of America Series. Charleston, SC: Arcadia Publishing, 2005.

Toler, William Pennington Toler, and Harmon De Pau Nutting. *New Harlem Past and Present*. New York: New Harlem Publishing, 1903.

Torrey, Raymond H., Frank Place Jr., and Robert L. Dickinson. *New York Walk Book*. 3rd ed. New York: American Geographical Society, 1951.

Turco, Peggy. *Walks and Rambles in the Western Hudson Valley*. Woodstock, VT: Backcountry Publications, 1996.

Turco, Peggy, and Katherine S. Anderson. *Walks and Rambles in Westchester and Fairfield Counties: A Nature Lover's Guide to 36 Parks and Sanctuaries*. Woodstock, VT: Backcountry Publications, 1993.

Twomey, Bill. *East Bronx: East of the Bronx River*. Images of America Series. Charleston, SC: Arcadia Publishing, 1999.

Ultan, Lloyd. *The Northern Borough: A History of the Bronx*. Bronx, NY: Bronx Historical Society, 2009.

Van Valen, J. M. *History of Bergen County, New Jersey*. New York: New Jersey Publishing and Engraving, 1900.

Ver Nooy, Amy. "The Ghost at Fiddler's Bridge and Other Spooks." *Year Book: Dutchess County Historical Society* 42 (1957).

Village of Suffern Bicentennial Committee. *Suffern: 200 Years, 1773–1973*. Suffern, NY: Village of Suffern Bicentennial Committee, 1973.

Voelbel, Margaret M. *The Story of an Island: The Geology and Geography of Long Island*. Point Washington, NY: Ira J. Friedman, 1965.

Walters, George R. *Early Man in Orange County, New York*. Middletown, NY: Historical Society of Middletown and the Wallkill Precinct, 1973.

Walton, Frank L. *The Cedar Knolls*. Bronxville, NY: Cedar Knolls Colony, 1960.

———. *Pillars of Yonkers*. New York: Stratford House, 1951.

———. "Sigghes Rock." *Westchester Historian: The Quarterly of the Westchester County Historical Society* 42, no. 3 (Summer 1966).

Ward, Seeley E. "Recollections of the Sloatsburg Area." *Orange County Historical Society Publication*, no. 2. (1972–1973).

Warner, Fred C. "Lady of the Cave." *Westchester Historian: The Quarterly of the Westchester County Historical Society* 40, no. 3 (1964).

———. "North Salem's Great Boulder." *Westchester Historian: The Quarterly of the Westchester County Historical Society* 32, no. 1 (1956).

Washington, Erik K. *Manhattanville: Old Heart of West Harlem*. Images of America Series. Charleston, SC: Arcadia Publishing, 2002.

Westchester Historical Society. *Anne Hutchinson and Other Papers*. White Plains, NY: Westchester County Historical Society, 1929.

Wilcox, Stanley, and H. W. Van Loan. *The Hudson from Troy to the Battery*. Philmont, NY: Riverview Publishing, 2011.

Wingerson, Roberta. "The Hunter Brook Rockshelter." *The Bulletin: New York Archaeological Association*, no. 68 (November 1976).

Wyckoff, Jerome. *Rock Scenery of the Hudson Highlands and Palisades*. Glens Falls, NY: Adirondack Mountain Club, 1971.

Zimmerman, Linda, ed. *Rockland County Century of History*. New City, NY: Historical Society of Rockland County, 2002.

INDEX

S

Y

Z